Killing the Hidden Waters

Killing the Hidden Waters

by Charles Bowden

University of Texas Press, Austin

Requests for permission to reproduce material from this work
should be sent to Permissions, University of Texas Press, Box
7819, Austin, Texas 78713.

Library of Congress Cataloging in Publication Data

Bowden, Charles.
  Killing the hidden waters.
  Bibliography: p.
  Includes index.
  1. Water, Underground—The West.
  2. Arid regions—The West.
  3. Indians of North America—The West—Economic con-
ditions. I. Title.
GB1019.B68   333.9'104'0978   77-5633
ISBN 0-292-74306-8

Photographs by William Dinwiddie; W. J. McGee expedi-
tion, 1894. Courtesy of Smithsonian Institution, National
Anthropological Archives.

Grateful acknowledgment is made for permission to publish
the following poetry: p. 58 reprinted from R. M. Underhill,
Social Organization of the Papago Indians, New York: Columbia
University Press, 1939; pp. 66, 69 by permission from
Piman Shamanism and Staying Sickness, Donald M. Bahr, Juan
Gregorio, David Lopez, Albert Alvarez, Tucson: University
of Arizona Press, copyright 1974; pp. 9, 65, 66, 67, 68,
*Singing for Power: The Song Magic of the Papago Indians of South-
ern Arizona,* copyright © 1969 by Ruth Murray Underhill
reprinted by permission of the University of California Press.
Reprint of 1938 ed.

94212

For

Jude Bowden, 1896–1975.
He never gave up or gave in.
But he gave all the same.

Contents

MAPS

Thanks

I've had a lot of help. The essay has been read by Kathy Dann-reuther, Zada Edgar, Richard Felger, Nancy Ferguson, Bernard Fontana, Lewis Kreinberg, Barbara Kreinberg, Gary Nabhan, Lawrence Clark Powell, and William Appleman Williams. They all tried to make it better.

I got interested in the desert and groundwater problems while working at the Office of Arid Lands Studies, University of Arizona, Tucson. Their support is in a class by itself.

Dan Matson, a linguist at the University of Arizona and a native of these parts, tried to explain to me the rhythm of life in the Sonoran desert before the post–World-War-II boom. His contributions are unfootnoted and omnipresent in the text.

Lawrence Clark Powell of the Graduate Library School, University of Arizona, read the essay, encouraged me, and did the one thing I could never do on my own: find a publisher. I'm grateful.

In early 1975, I spent a few days with ecologist H.T. Odum. His work is seldom cited in this essay but it was constantly present in my mind. He pushed the rhetoric of ecology past metaphor and into palpable reality for me. I hope he will forgive what I've done to his song.

There would be no book without Bernard Fontana of the Arizona State Museum. He shared his vast knowledge of the Sonoran desert and its people. And he taught me. Without his help, advice, and criticism, I could not have done this essay.

Finally, I'd like to thank the late J. Newfoundland. He didn't give a damn about this book, and his healthy perspective on such matters saved me from completely losing my own.

Pima Indian holds a calendar stick.
Smithsonian Institution Photo No. 2693-A.

Part One

Long ago, they say, when the earth was not yet finished, darkness lay upon the water and they rubbed against each other. The sound they made was like the sound at the edges of a pond.

—Papago story of the creation

Behold, I will do a new thing; now it shall spring forth; shall ye know it? I will even make a way in the wilderness, and rivers in the desert. The beasts of the field shall honor me, the dragons and the owls: because I give waters in the wilderness and rivers in the desert, to give drink to my people, my chosen.

—Isaiah, 43:19–20

1. The lay of the land

The days and nights slash across the Covered Wells stick. This five-to six-foot piece of saguaro rib holds gouges and nicks. They are the marks of the Papago, a Piman people, and testify to their effort to discover what time and the Sonoran desert meant in their lives. The Papago, also known as the Papavi, the bean people, the desert O-otam, lived in and with one of the most arid stretches of North America. For centuries, they survived the land's heat, the land's rock, the land's three to ten inches of rainfall, the land's wild swings of want and abundance. Early in the nineteenth century they began notching the sticks, and history intruded into their minds with its love of the particular, the taste, the feel, the cost. The sticks would be handed down from man to man, generation to generation, with gouge and nick a clue to memory for what mattered among all the things that had happened to the people.

The Covered Wells stick remembers when groundwater came to the people through the white man's pumps. It links a world based on renewable resources with a world based on mining nonrenewable resources. In arid lands groundwater is essentially like coal or oil: easy to exhaust, hard to replace. The story cut into the stick is an examination of what groundwater means to people and what it can do to their ways and means. The desert people knew that water from the earth was energy, and that like all forms of energy it must either be mastered or it will master. Water could never be just water in this place because here it determined both the amount and kind of life to be lived.

Most whites that hear the tales in the desert Indians' sticks are jarred by what they contain and by what they find beneath notice. Mainly, they find white people beneath notice. The Covered Wells stick does not recall the Mexican War. It has no mark for the gold rush of 1849, when tens of thousands of Americans on their way to the west coast traversed the lands of the Piman people. Sometimes the stick recalls nothing at all. Take 1845: "In that year nothing happened worthy of mention." Three years later there is silence on the Mexican and American contest for the Southwest and lengthy mention of a snowstorm. While half a million Amer-

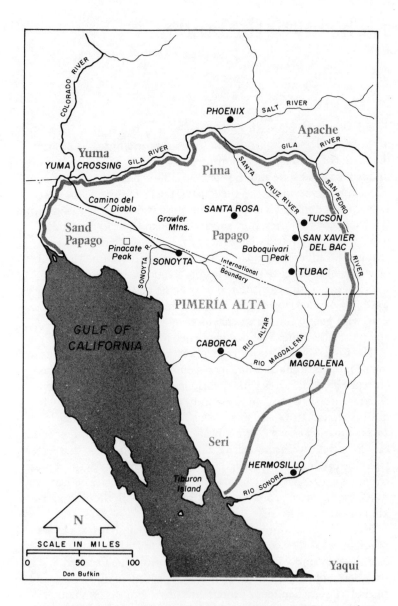

MAP 1. Pimería Alta. By permission from *Friars, Soldiers, and Reformers: Hispanic Arizona and the Sonora Mission Frontier, 1767–1856,* John L. Kessell, Tucson: University of Arizona Press, copyright © 1976. Redrawn by Don Bufkin.

ican boys go to their graves wearing blue and gray, the stick focuses on tribal games, gambling, the fact that two bachelor councilmen are given wives. Whites tend to make the stick because of their machines, and the machines rate a notch because they kill.

> 1879—During this year the Southern Pacific Railroad came from California . . . to Casa Grande.
> 1880—A Papago was run over and killed by a train in Casa Grande. (Kilcrease, A.T., 1939)

The pattern repeats. A mine opens in Quijotoa in 1882; the stick says that in 1885 a faulty blast killed an Indian working in the mine (ibid.).

Mainly the record stays with the doings of an Indian world. In 1893 when American historian Frederick Jackson Turner explained the end of the frontier to the citizens of the republic, the desert people were struck by different news.

> The Papago women have always played a game called "Tauka." The players manipulate two rawhide balls, about one and one-half inches in diameter and held together about an inch apart by rawhide thongs, with mesquite sticks about five feet long and shaped like a shepherd's crook. The rules are similar to those of football.
>
> This year the women began to bet on this game with the wildest abandon. It is indicated that some, after losing the property that several years of prosperity had brought them, bet their clothes. This caused a great deal of comment among the Indians and was deemed worthy of note in this history.
>
> The rain feast was celebrated in all its splendor. (Ibid.)

And that is all for 1893.

The largest entry in the Covered Wells stick recounts the drilling of a well in Santa Rosa village in 1912. This was the work of white men employed by the United States government. The Americans who put in the well represented a people who had crossed the continent from the humid East to the arid West with practices and machines that devoured large amounts of water. When past the one-hundredth meridian the rains had grown scant, these people had learned how to punch holes into the earth to reach the groundwater and how to fashion machines to lift it. By these acts they hoped to insure that their ways of living could continue regardless of the place where they put down their roots. By 1912 white culture had mastered the windmill and was on the verge of wide-scale use of oil-fired pumps.

For the Americans it was inconceivable that a well in a desert village could be anything other than a boon. Considering the hard facts of Papago life, the well in Santa Rosa village must have seemed a greater boon than most. The stick describes the daily struggle of the people to get water. During the brief rainy season there was plenty in the wash. But when the wash went dry, the women carried water in jugs from a small spring eight miles away. They would leave early in the morning and return during the heat of the day. These were the people who fought the well.

The cost of the improvement would be charged to the tribe, and this was the first reason given against it. The chief called together his advisors and they talked and they unanimously rejected the well. It was not needed: "For although we do not have to pay for the well now, sooner or later the money must come. The people have lived a long time on their lands and prospered without this improvement and they wish to continue to live without the gratuitous assistance of anyone." The white men listened and then they took the chief to an anthill and shoved a twig into the ground. They said, "Here a well shall be drilled." The stick reports, "It was done" (ibid.).

Presented with a hole in the earth that made the sixteen-mile trek with ollas unnecessary, the people offered a second reason against the well. The chief asked that the thing be ignored. Water must be lugged for miles in the heat "in order that the Papagos might continue their old life . . . ."

This was the heart of the matter. The Papagos were a desert people and they had survived by learning to live with seasons, with two villages, with the harvest of desert plants, with the resources of their land. The well lanced into the sinews that bound man to man, family to family, band to band. But the well was too mighty, the energy too great. "Gradually, however," continues the stick, "but reluctantly (for the Papagos are unusually law abiding) they began to disobey and get their water from the well" (ibid.). The chief threatened and then beseeched them to desist. One night he was thirsty and there was no water in his home. He stole with his olla to the well. He was caught. The well had come to Santa Rosa village.

What the wells did (and they continued to be drilled by the
Bureau of Indian Affairs for decades) was help transform the
Papago from a people who had lived off the land to a people who
happened to live on the land. The desert O-otam moved from a
society of abundance to a society of privation. Today they are
among the poorest reservation Indians in the United States. A
tribe once obsessed with self-sufficiency and mutual aid collapsed
into a world where jobs come from alien cities, energy comes from
fossil fuels, water comes from holes in the ground that pour forth
millennia of rainfall in a man's brief three score and ten.

The Papagos, ignorant of hydrology, bereft before the white
man of so much as a wheel, know in their hearts and lives a lesson
just becoming apparent to Americans. Water is energy, and in arid
lands it rearranges humans and human ways and human appetites
around its flow. Groundwater is a nonrenewable source of such
energy. These facts are the core of the impact when groundwater
is developed in such places. Humans build their societies around
consumption of fossil water long buried in the earth, and these
societies, being based on a temporary resource, face the problem of
being temporary themselves.

To ask what is the impact of developing groundwater in arid
lands is simply to seek the price that must be paid for this unique
human knack of influencing the availability of water. The answer
is this: man builds water-rich societies in arid lands by living out
of balance with his water supplies. He uses water faster than it
can be replaced by rain. When this fact becomes obvious, people
call it the groundwater problem.

What man tinkers with is a tiny part of a huge hydrosphere.
Estimates place the total amount of water girdling the earth at
around 1,500 km³—visually a depth of 3,000 meters spread evenly
over the globe. Only about 5 percent of this moisture is fresh, and
75 percent of this small portion is locked away from man in snow,
ice, or permafrost. Humans can reach about 1 percent of the total
hydrosphere. Some of this accessible water is in vegetation, rivers,
and lakes. But the bulk, 99 percent, is in the ground at depths of
less than 1,000 meters: groundwater (Dooge, J.C.I., 1973).

This trifle is the basis of the water laws, the water conferences,
the civilizations. The groundwater part of the hydrosphere was
largely neglected by humans until the recent past because it was
awkward and expensive to lift to the surface.[1] In the last century,

cheap energy and growing human numbers have steadily increased the pumping of buried water; there is some evidence that this effort is raising the level of the oceans a bit (ibid.).

The Papagos, or desert O-otam, are a people who once lived completely in balance with their water resources and now live out of balance with them. To look at their history is to look at both sides of the coin. Veterans of life in an arid land, they have produced a culture highly sensitive (at least until quite recently) to the preciousness of water in a desert. When in the 1930s whites attacked the drunkenness of their annual rain festival with its massive consumption of cactus wine, one old Papago said angrily, "Do the whites not understand that we have no water except what comes from the sky? We have no canned food, so we need the corn to feed our children. We have no automobiles so we need hay for our horses. Why then do they say we should not drink the cactus liquor?" (Underhill, R.M., 1938). The desert teaches reverence for water; in the Middle East, rainfall trapped for drinking purposes is called "the water of God's mercy" (Downing, T.E., and M. Gibson, eds., 1974).

The desert O-otam, seeing moving water so seldom, thought about it a great deal. A chant sung during their annual salt trek to the Gulf of California shows that they sensed the nature of the hydrologic cycle.

> By the sandy water I breathe in the odor of the sea,
> From there the wind comes and blows over the world.
> By the sandy water I breathe in the odor of the sea,
> From there clouds come and rain falls over the world.
> (Densmore, F., 1929)

This was a great insight for a desert people so short of water that their art and architecture were dry.

Their lands, Papaguería, lacked a lake, river, or brook. Warriors proved their courage by wading into the Gulf of California and letting the waves lap against their chins.

Papago life was premised on the fact that water was constantly in motion and that humans could endure only by fitting into this movement. The lesson was written across the landscape of the Sonoran desert. Vegetation clung to the arroyos and ephemeral rivers. As one moved off these topographical concentrations of water, life thinned with the dilution of moisture. A man could

walk from alpine forest to cactus bajada in a day and witness a
succession of floral worlds separated by gradations of rainfall. The
teachings of the land left their mark. The language of the people
was saturated with words derived from one thought: water. The
great rituals of the O-otam focused on songs to pull down the
clouds.

> The sun children
> Are running westward
> Hand in hand
> Madly singing,
> Running.
>   (Underhill, R.M., 1938)

To the south the rains fell away and the desert grew inhumanly
dry. For the Papago this was "the direction of suffering" (ibid.).

For modern industrial man this Indian water world is both
simpleminded and haunting. Today the hydrologic cycle has escaped song and been broken into disciplines, categories, books.
Papago observations seem elementary. But no amount of information can change the facts of the hydrologic cycle, and that is what
haunts the new lords of the American Southwest. Where the
O-otam sang for power, moderns dam rivers, mine coal, pump
petroleum. The need for power is the link between the two peoples
and the two ways. The rain does not come to land often enough;
the rivers do not run full enough; the aquifers do not burst to the
surface easily enough. Man must invest his time and energy. And
then, this done, the groundwater does not last long enough.

A good way to realize some of the constants in arid lands is to
look at some of the problems Americans have confronted in exploiting groundwater. First, they have learned that water has an
energy value. The sun lifts it from the seas (evaporation) and
moves it across the land (wind). Temperature variations expand
or reduce the ability of the atmosphere to hold it (rain). Having
been lifted to great heights, the water has power as it flows back
to the oceans. It can wear mountains, erode soils, carve canyons,
spin turbines, flood land. Because rainwater is almost pure, it has
a great potential energy for interacting with other molecules.
Pumped through a plant by the sun, it can help capture solar energy
(photosynthesis). Run it across deposits of salt and it will combine
with salt (salinity). Run it through a factory and it will remove

material with its driving power or chemically combine with material and thus cleanse. The more it reacts with other elements the dirtier it gets and the less potential energy it contains. For this reason humans will pay a premium for pure water. Returned to the sea, the water is recharged by the sun, and the cycle continues— without beginning or end, a flow.

These simple facts about the energy value of water are a key part of what has come to be called the groundwater problem in arid lands. First, the inevitable nature of groundwater in such regions: it is essentially a nonrenewable resource. Then, once the decision has been made to pump this easily depletable resource, the crude truths of energy assert their importance. Take the ability of water to combine with other materials. What this means is that, left in the ground for millennia (a common occurrence in the fossil deposits of arid regions), water molecules combine with the minerals of the earth. The result is that desert aquifers are often of low quality—brackish, saline, hard. The way to alter this condition is by adding energy. Humans call this desalination or filtration. But this expenditure to change the quality of water is a fixed cost of doing business. Another inescapable cost is the price of lifting the water to the surface. Technology can change the source for the lift; the lift remains forever.

Once the water is lifted to the surface, there are still more bills to pay for using groundwater in such regions. Sunlight can alter water from a fluid to a gas. But solar energy is inadequate to vaporize the minerals in the water. In arid land irrigation systems this has led to a worldwide problem called salinity. The sun evaporates the water and leaves the salt. Since humans by a massive energy investment have concentrated huge amounts of water on small plots of earth (irrigated agriculture), they have also concentrated huge amounts of salt. Left untended the fields go sterile.

Dealing with salinity calls for large investments of energy. Driving the salt down below the root zone of cultivated plants calls for about twenty percent more water being delivered to the field. This solution is expensive and short term. The salt driven down concentrates in the groundwater, making it increasingly saline. This impasse can be avoided by installing drainage pipes (yet one more investment) to carry salt-laden water from the field. But this salt water goes somewhere. In the Colorado basin of the United States it went to Mexico and ruined farming land. Now a treaty is

forcing the United States to pay for a large desalination project at
the border.

Waterlogging is another woe attendant upon using groundwater.
After being pumped to the surface, the water sometimes fails to
sink into the earth at a sufficiently fast rate. The desert field becomes a local swamp. The solution is to put in more pumps driven
by more energy to remove the water. Having paid the toll charge
for salinity and waterlogging, another expense confronts exploitation of desert aquifers: subsidence. It is common parlance to state
that groundwater left in the ground is out of the hydrological cycle
and pretty much useless to man. When this water is pumped from
the desert basins the bit of work it was accomplishing sometimes
becomes obvious. It was holding up the earth. Drained of water,
the soil contracts and the land sinks in an irregular and tearing
fashion. Mexico City is a good example; huge buildings have
cracked as they sank into the earth. Gas lines may break. Homes
may shift apart.

These facts of depletion, salinity, waterlogging, and subsidence
are usually considered self-evident, elementary, and not worth
talking about. Since they are not worth talking about they are often
ignored. That is why, when they happen, when they reassert the
inevitableness of their presence, they are hailed as problems, and
conferences, studies, and scholars are hurled at them. They are
treated as ugly surprises. Clearly, they are not. They are part and
parcel of the energy potential of water, of the hydrologic cycle, of
the nature of aridity. Like the moral rules taught the young of
human beings, these facts of water and deserts form a touchstone
that must be constantly referred to—not because they change, but
because they never change.

The most interesting facet of the impact of groundwater development in arid lands is variable: human use of groundwater and
human response to that use. The planet holds a legion of cultures
in the dry zones who have confronted the deserts, the steppes, the
rainless mountains. These various groups of people have used
resources at different rates with different results. Often this record
of human activity in arid lands is obscured by such pufferies as
tributes to human ingenuity or celebrations of a sense of motion
called progress or loose talk under the heading of the conquest of
nature. Underneath this skin of language a basic process is always
going on: people are using different devices and forms of organiza-

tion to influence the flow of materials and the concentrations of materials at specific sites.

Why resource use varies from people to people around the globe is not really understood. Modern America is built on the same terrain as prehistoric America. The aborigines had even more coal, uranium, iron, and petroleum at high concentrations than the current occupants of this land. But the aborigines did not use these storages and endured a world driven by solar energy.

For some reason or reasons humans at various times and places have successfully struggled to increase the flow of materials toward themselves. An example of how complicated this question can be is offered by the Yir Yoront of the Australian bush. This group lives on the coast, and their only mode of water transport is a log which they cling to. The waters they traverse are infested with sharks, crocodiles, stingrays, and Portuguese man-of-wars. The Yir Yoront are frequently stung, bitten, and devoured. Forty-five miles north of them is another group of bushmen who use bark canoes. The Yir Yoront know them and have the materials to make similar vessels. They do not do this because, they say, their ancestors in the myth world that preceded time and pain lacked canoes, and thus, they are fated to lack them (Spicer, E.H., ed., 1952).

The cultural inertia that grips the Yir Yoront is fast disappearing from the arid lands. By adding power from fossil sources, man is swiftly increasing the rate of flow for water and other materials. Population groups that remain locked into older ways tied to solar flows are being pushed aside or extirpated. Resources are being consumed at faster rates through the addition of power (coal, gas, petroleum, et al.)[2]

Given the blunt facts of water in arid regions (salinity, depletion, waterlogging, subsidence), some things considered problems can be dismissed as the product of the basic mechanics of exploiting water resources in the drylands. But other parts of what is considered the impact remain intriguing. Groundwater is being depleted in arid areas by the founding of cities, and by the expansion of modern, mechanized agriculture. If groundwater is basically fossil and nonrenewable in arid regions, then one asks whether it should be used at all, and if used, for what?

So the first thing to do is to look at the plans, the place, and the people. There is more to groundwater than the fact that two atoms of hydrogen and one of oxygen sometimes occur at depth. The Covered Wells stick suggests there is much more.

Yield

The bulk of the groundwater depletion in arid lands has been for agriculture. On a planet of limited arable land, limited fossil fuels, and limited groundwater, a growing global population is the reason for the mining of ancient aquifers. It is argued that by exhausting nonrenewable resources at a faster rate and on a grander scale, famine will be averted and the lot of human beings improved. Sometimes this practice is called the green revolution, sometimes making the deserts bloom, and sometimes the miracle of modern agriculture.

The machines do not change the hungers. A plant requires water, fertilizer, air, soil—six basic elements: hydrogen, oxygen, nitrogen, potassium, phosphorus, sulphur. Man can work with a stone hoe or a twenty-thousand-dollar tractor, still these needs are fixed and immutable. What man can do is influence the amount of these materials that are present, the place that they are present, and the kinds of plants that receive them. This is called agriculture, the concentration of resources through energy expenditure and the creation of structure. The greatest strides in moving the materials essential for photosynthesis have been made in the western industrialized nations, particularly in the United States.

This nation has learned how to replace men with machines, manure with chemical fertilizer, rain with irrigation, insects with pesticides, weeds with herbicides, traditional crops with hybrids, hunger with plenty. With four billion humans alive today and a world population pegged at seven billion by the year 2000, the American way of industrial farming has aroused interest and hope. Mounted on machines a citizen of the republic can raise a hectare of corn with only 22 man-hours of labor. A Mexican campesino toils 1,114 man-hours to achieve the same end (Pimental, D., et al., 1975). Since the arid lands are largely empty, sunny, and amenable to mechanized agriculture, it is thought that they offer fertile ground for the extension of western food-raising technology. There is talk of conquering the deserts.

This talk has hit some snags. While the planet's resources of minerals and of water cannot be increased or diminished appreciably, the location and concentration of these materials can be. At present rates of growth and consumption, bulk deposits of various metals and fuels and of groundwater will be seriously

depleted in coming decades.[3] The ingenious machines that lift
water to the irrigation districts are sucking down the water table
at rates that vary in all but their ultimate conclusion. The natural
gas that powers the pumps is giving out. The fossil fuels driving
the fertilizer factories are giving out. The tractors, herbicides, insecticides, the army of equipment, drink petroleum faster than it
can be found. Development in modern usage means expanding
these appetites from the industrialized West to those portions of
the globe currently practicing methods less consumptive of resources. Development means converting the entire planet to the
use of fossil water, fossil fuels, and diminishing ore deposits.
When some humans from time to time point out this fact they are
told that any problems encountered in the course of development
will be overcome by a thing called the technological fix. Increasingly this fix is the thing upon which it seems billions of human
lives will depend.

To look at the land is to see the dimensions of the problem and
the wall blocking an easy solution. About 3.4 billion acres are
currently farmed globally; some think another 3.4 billion could be
pressed into production (Wittwer, S.H., 1975). Presently, the
earth is being exploited in the following manner: 11 percent is
farmable and most of this is under the plow; 22 percent is devoted
to livestock in the form of meadows, ranges, and pastures; forests
cover 30 percent; the remaining 37 percent is considered unsuitable
for agriculture because it is too cold, dry, or steep (Pimental, D.,
et al., 1975). Obviously, these numbers invite two tactics. The
amount of cultivated land can be expanded by diminishing other
forms of land use, or land already cultivated can be more intensively farmed.

Both tactics increase the use of exhaustible resources. To farm
more intensively means to farm more frequently and with a greater
investment of labor and materials. It means tilling more often,
fertilizing more often, watering more often, spraying chemicals
more often. Such methods produce food and diminish the ability
to produce food in the future. In the United States this has been
the path chosen in past decades. Americans are mining the soil itself. For example, Iowa corn production consumes about 36 metric
tons of topsoil annually per hectare. Nationwide, about 3.6 billion metric tons of topsoil are lost annually—this runs around 31
metric tons per hectare of cultivated land. If the nation chose to
replace this annual loss, it would have to let its farmland lie fallow

for eleven years. American agriculture is using the earth at a rate eleven times faster than it is naturally replaced (ibid.). This feat of legerdemain is possible because Americans are compensating for the vanishing topsoil by pouring fertilizer onto the land. This fertilizer is either derived from fossil fuels, captured and delivered by fossil fuels, or both. One exhaustible resource is being replaced by another exhaustible resource.

The other tactic, that of expanding the amount of land under cultivation, poses more of the same problems. Land currently innocent of the plow is usually land either too dry, or too cold, or too steep for easy exploitation. To bring such ground into production means altering undesirable features by investing resources. In arid lands, this comes down to irrigation to escape undependable rainfall and the use of fertilizer to ensure multiple cropping and compensate for the soil's lack of organic content. Generally, development in arid lands means mechanized, irrigated agriculture. Today, 15 percent of the world's cultivated land is irrigated. This man-watered land produces 30 percent of the world's food. Most of it is in mainland China, the USSR, the U.S.A., Pakistan, and India. It represents a tremendous commitment of energy; irrigation simply means moving water by contrivance and pump instead of letting the sun deliver it as rain. The high yields result from high investments of water, energy, fertilizer. Irrigated agriculture is energy intensive, perhaps the paradigm of modern agriculture.

India and the United States have about the same amount of ground under cultivation, 350 million acres. India produces only two-fifths as much from this base as the United States. The difference is fertilizer, water, herbicides, pesticides, and machines. (Wittwer, S.H., 1975). This is what the green revolution means. In seeking to exploit the new miracle rices and miracle wheats, India increased fertilizer use ninefold (from 300,000 tons to 2.8 million nutrient tons) between 1960 and 1974. In the six years following 1965, the number of diesel and electric pumps in the nation rose from 979,000 to 2.7 million; private tubewells jumped from 113,000 to 550,000 (Gavan, J.D., and J.A. Dixon, 1975). High investments of water and energy resulted in high yields.

The plan is to push this type of agriculture (Stakman, E.C., R. Bradfield, and P.C. Mangelsdorf, 1967). There are 40 countries in the world where per capita income is below $150 per annum. They are targets. There is a loose assemblage of hunger and large families called the third world. They are targets. The American

AID program seeks to improve "agricultural technology" in these populations by agricultural research, extending the green revolution, expanding dryland farming, and inducing small farmers to undertake "high productivity agriculture" (Agency for International Development, 1975). These visions of solving food shortages and fossil fuel shortages are not easily disturbed.

To understand the plight of modern agriculture one must take a good look at the ingredients of this method. A key element is fertilizer lavishly expended upon the crops. "The natural supply of fixed nitrogen is limited and it imposes a limit on the capacity of world agriculture (Safrany, D.R., 1974). About 79 percent of dry air is nitrogen, but to make this vital element available to plants energy must be spent and nitrogen joined with some other element. Legumes achieve this result by using solar energy. However, the rate is slow and often requires fallowing the land. Man has escaped this constraint by technology (now about sixty years old and originally perfected to facilitate the production of ammunition). By consuming natural gas or some other fuel, nitrogen is taken from the atmosphere and combined into ammonia. This energy cost, now paid with fossil fuels, is the rub. "For this reason," concluded one observer, "any large reduction in the cost of fixed nitrogen is likely to result not from improvements in the technology of fixation itself, which is already highly developed, but from advances in the technologies that supply the energy or raw materials needed for this fixation" (ibid.). At present, American production of ammonia is doubling every five years, and the fossil fuels used in this production are dwindling.

The problem continues right across the board. Fertilizer, already growing scarce and more expensive, accounts for 30 to 40 percent of the increase in productivity achieved by American agriculture in recent decades. Gains in the underdeveloped world are about half due to the higher use of fertilizer. Generous use of pesticides explains about 25 percent of the United States output of crops, livestock, and lumber; their abandonment would raise farm prices 50 percent (Wittwer, S.H., 1975; Ennis, Jr., W.B., et al., 1975). The mark of exhaustible materials can be seen in the pumps, tractors, and methods of modern farming (Poleman, T.T., 1975; Paddock, W.C., 1970). With a planet facing a food deficit of from 55 to 85 million tons per annum in the developing countries, these facts are sometimes ignored. But they keep popping up again. "The Less Developed Countries," one critic notes, "are not going to have much leverage on world fertilizer prices over the next decade . . . and they will have even less on energy prices, also significant because of the importance of petroleum in fueling tractors and driving irrigation pumps" (Crosson, P.R., 1975). The reasons for this lack of leverage are simple. Today, 86 percent of the world's fertilizer, for example, is used in the developed countries, which contain only 39 percent of the world's population (Wittwer, S.H., 1975). As energy costs rise and concentrations of resources dwindle, the poor nations will have to compete with this collection of states already devouring a lion's share of earth's bounty.

There is one further twist to this tangle of declining concentrated resources and rising resource appetites. In the past decade a debate has erupted over who makes the most efficient use of resources, the developed or the underdeveloped world. Energy expended in producing a crop is being compared with the energy reaped from the crop.[4] Results of these analyses vary. Some investigators find modern agriculture operating at an energy deficit (Steinhart, J.S., and C.E. Steinhart, 1974); some find a slight yield (Pimental, D., et al., 1973); and some find a good yield in the field (Wittwer, S.H., 1975).

But even the most optimistic assessment gives cause for alarm. Wittwer concludes that for every calorie of energy invested in cereals and legumes by modern agriculture, this mechanized method netted three to five calories of "food and feed energy" in the field. However, energy must be spent to get the produce from the cropland to consumer. "It is estimated," he continues, "that our present total food system expends at least five units of fossil energy for each unit of food energy produced" (ibid.). Americans are eating coal, natural gas, and petroleum. Their agricultural system is based on extravagant use of these resources.

As these fuels decline, the system is placed in greater jeopardy. Look at it this way: a nation of two hundred-odd million humans is feeding itself and millions of other people not simply by capturing the renewable cascades of sunlight, but by exhausting fossil fuel supplies in this effort. When these storages of energy are gone (a matter of decades for some of them), the miracle of American agriculture, the big yields, will end because there never really was a miracle—there was just the profligate use of fossil energy in farming.

Societies lacking access to these fossil storages cannot use them.

Instead such societies must depend upon human labor, the natural delivery of water, or fallow; they must contend with losses to insects and low yields per acre. Since these groups cannot lay hands on abundant reserves of energy, they cannot lavish many calories on their crops. They endure a state of positive yield: more calories are reaped from a crop than the farmer personally invests in it.

There are men in the hills of New Guinea who live in a stone age. They have a better yield on a per-calorie-invested basis than the American farmer (Rappaport, R.A., 1967). There are men in the jungles of Latin America who slash and burn their land, farm, and then move on to slash and burn again. They have a better net yield than the American farmer. Similar men use digging sticks in Africa and toil as hungry millions in India and across Asia (Odum, H.T., 1971; Steinhart, J.S., and C.E. Steinhart, 1974). They use no subsidy from energy resources laid down millions of years ago as petroleum, natural gas, coal, uranium. They live in a solar-based economy, and in such an economy, negative yield often means death. If a man poured more calories from his body in raising a crop than he could reap from the crop, he would need no elaborate calculus to compute the energy flow. He would starve to death.

The reason there has been a debate of recent years over the net yield of modern agriculture is that some people find the significance of such measurements difficult to grasp. Their importance can be stated rather simply:

> What a sad joke that a man from an industrial-agricultural region goes to an underdeveloped country to advise on improving agriculture. The only possible advice he is capable of giving from his experience is to tell the underdeveloped country to tap the nearest industrialized culture and set up another zone of fossil-fuel agriculture. As long as that country does not have the industrial fuel input, the advice should come the other way.
>
> The citizen of the industrialized country thinks that he can look down upon the system of man, animals, and subsistence agriculture that provides some living from an acre or two in India when the monsoon rains are favorable. Yet if fossil and nuclear fuels were cut off, we would have to recruit farmers from India and other underdeveloped countries to show the now affluent citizens how to survive on the land while the

population was being reduced a hundredfold to make it possible. (Odum, H.T., 1971)

Industrialized societies have decided that the important criterion in measuring farming productivity is the yield of energy and not the investment. Progress is determined by raising a field from 40 bushels of something to 80 bushels of something. What goes into making this increase is given scant attention. Thus, Americans in general seem unaware that behind the glories of their agriculture lie a standing army of plant scientists and extension agents, a gluttonous appetite for fossil fuels, and legions of machines. Crop varieties are lauded if they prove lusty consumers of fertilizer and water. This way of thinking makes perfect sense in a world where coal, oil, natural gas, and groundwater are limitless, renewable resources.

It is a pervasive mode of thought. Economies are geared toward growth, and the health of an economy is considered improved if over the course of a year it manages to find ways to consume resources at a faster rate. A leading volume on development of the third world, Esther Boserup's *The Conditions of Agricultural Growth* (1965), echoes this line of logic. Boserup argues that population growth is desirable in such countries because it pushes humans past the resource rate of solar-based economy and forces them to join the West in devouring storages of minerals and fuels lest they starve.

Over the past few decades, this appetite for imbalance with resources has penetrated thinking on groundwater. Once hydrology boasted a concept called safe yield (Kazmann, R.G., 1966; Mann, J.F., 1961; Harshbarger, J.W., 1971). It meant that aquifers should be pumped no faster than they are naturally recharged. Pursuit of this notion seriously limited what humans could achieve in arid lands. For in these areas water is scarce and that scarcity is the factor which constrains life.

The vegetation reflects this brutal dryness. The succulent cacti store water like tanks and towers as insurance against the months without rain. Some trees, like the mesquite, search the earth at depths up to 175 feet looking for moisture. Annuals in such regions are like invading hordes that flourish during the brief periods of rainfall, then must retreat into the hull of seed until the skies darken again. Human efforts in the arid lands are just as hampered by water availability. There is a valley in central Arizona with three quarters of a million people. There is a section of the Sonoran desert about the size of Connecticut with no human beings. Since the natural recharge to aquifers in arid regions is generally dishearteningly small, adherence to a policy of safe yield meant the acceptance of either a small society commensurate with the water supply, or of a large society with a very low per capita water use. It meant living in balance with water resources.

Such a policy might have a long future but it offered a limited present. "Experience has demonstrated," J.W. Harshbarger advises, "that bold decisions to utilize groundwater sources in excess of natural replenishment have proved to be a realistic water management method" (Harshbarger, J.W., 1971). In other words, safe yield as a concept has been abandoned. Harshbarger has traced the history of this retreat from safety. When the notion was first floated in the twenties it meant "the practicable rate of withdrawing water from it [an aquifer] perennially for human use" (ibid.). By the forties, this was amended with three conditions: use should not exceed recharge; lowering the water table should not lead to exorbitant pumping costs; and lowering the water table should not permit the intrusion of undesirable water (ibid). Within the last twenty years or so, the U.S. Geological Survey simply dropped the term. The American Society of Civil Engineers Committee on Groundwater revised safe yield to mean: ". . . the patterns of distribution and draft from wells in developed groundwater reservoirs commonly differ from the ideal pattern for yielding the maximum perennial supply. The safe yield may then depend upon the practicability of relocating wells rather than any characteristics of the natural resource" (ibid.).

Various reasons are given for abandoning safe yield as a concept. It is sometimes argued that safe yield is a pseudohydrological idea, meaning the idea of safety is a human intrusion into what should be hard science. Others say that by limiting pumping to recharge, humans are unable to use most of the aquifer and are restricted to just skimming the top. But the main reason given for fleeing safety is that the exhaustion of an aquifer may turn out to not be a problem at all. As H.E. Thomas put it in *Water and the Southwest—What Is the Future?*:

> . . . and several states have laws restricting development to safe yield. Such laws may provide satisfactory solutions to problems restricted to a single aquifer or to short range eco-

nomics. But in the broad view, if the water does not come
from storage within the aquifer it must deplete the supply
somewhere else in the integrated hydrologic system; on the
other hand, wholesale depletion may be economically feasible
in the long range view if it results in building up an economy
that can afford to pay for water from a more expensive source.
(Thomas, H.E., 1962)

Much of the arid world is, or is becoming, a laboratory for testing this idea of the benefits of mining aquifers. In America it has
been pursued rigorously in such spots as southern California, central Arizona, and west Texas. Now each of these regions is enjoy-
ing booming economies which rest on sinking water tables, and
each of them is actively seeking imported water from distant
points. Southern California has already reached out to the northern
part of the state and the Colorado River and is looking longingly
at the Pacific Northwest. Central Arizona is building a monster
pipe from the Colorado River and dreams of getting a share of the
Columbia River. West Texas wants an aqueduct from the Mississippi. The American West as a unit periodically floats the idea of
redirecting Canadian rivers toward the arid lands to the south. In
a similar fashion, the Russians want to turn Arctic rivers back
toward the heartland of their nation.

Obviously, this policy of vigorously exploiting groundwater is
based on the same kind of thinking as American agriculture as a
whole. In each instance current prosperity is based on using resources at a rate greater than their natural replacement. Water,
petroleum, topsoil, and so forth, are all spent as if there were no
tomorrow because it is argued that only through such practices can
humans create an economy healthy enough to face tomorrow.
When tomorrow finally comes, it is argued, humans will have
found a technological fix or they will be so prosperous that they
will import water from somewhere and gladly pay the price.

In both instances when the price finally comes it is high. Declining fossil fuel stocks are swiftly increasing the cost of American
agriculture and the charge for its products. Imported water, when
it arrives in spots like southern California and central Arizona, is
expensive because it must be pumped uphill. And this pumping is
done with fossil fuels, which, as they decline, increase in price.
High productivity in both cases is achieved by high investments.

The arid lands are a merciless place for such societal experiments. They are full of dead people and cultures because the weather fluctuates a great deal. For aridity is more than just dryness; it is guaranteed uncertainty. Humans in the past and in the present have sought to overcome these oscillations by finding storages and concentrations of soil, timber, and water. In the Middle East, north Africa, parts of Asia, and the American Southwest, these efforts have been periodically rewarded by moments of great human density and prosperity, and then, for various causes all rooted in precariousness of climate, blasted. Eventually, in the past, the price of gathering water, food, and materials rose too high, and once-flourishing civilizations and Indian villages became tourist meccas and national monuments.

This past may be temporarily walled from the present by current tapping of fuel and water storages, but it lingers just outside that wall. Reminders come from time to time: the dust bowl of the American thirties; the mummified Sahelian dead of the seventies. Meanwhile, the arid world is racing toward that state called development. Arab governments surrounding the Persian Gulf are pouring petrodollars into the sand in order to create Western style farms and cities. Saudi Arabia, with possibly a five-decade petroleum reserve, and Iran, with maybe as little as fifteen years of dependable fuel, are casting off a solar-based economy for a fossil economy. Israel toils to modernize the Holy Land. In Libya deep wells are painting green circular fields on the terrain.

A planet of hunger is following the footsteps of Western technology; it is marching out of poverty wearing fossil boots. As fossil water, fossil fuels, and minerals in high concentrations decline, the industrialized West and struggling third and fourth worlds are demanding more of these resources. Where all this leads strikes some observers as self-evident: "With the human population projected to increase to 7 billion within 25 years and 16 billion by the year 2135, food shortages and energy, water and land limitations will become critical" (Pimental, D., et al., 1973).

While this resource problem is global, in the arid lands it becomes even more brutal. For in these sunny, dry regions, water cannot be depended upon. Heavy use of water eventually forces humans to use groundwater. This resource, unless carefully husbanded, is nonrenewable at any rate satisfactory to humans. Humans, by testing whether or not groundwater mining can be a springboard to a better world, toy with disaster in the deserts and semi-arid rangelands. Here the rains do not always come and the land can be very hard.

Place

The hunter, the farmer, the nomad can all find and know the arid lands. They can sense where they begin and end, and their lives become explanations of how arid lands differ from more temperate zones. Science has not been so fortunate. For decades it has wrestled with definitions that expand and contract such regions; various concepts of evaporation rates, minimum rainfall, rainfall spacing, vegetation types, and other yardsticks have been tossed around in fashioning criteria (Evenari, M., L. Shanan, and N. Tadmor, 1971; McGinnies, W.G., et al., 1968). Two belts of earth equidistant from the equator, the horse latitudes, are sectors of low rainfall. Depending upon the authority cited, anywhere from 15 to 20 percent of the globe's land mass is considered arid. Sometimes the poles are included because of their low precipitation rates.

The deserts and semi-arid rangelands are not so much occupied by unique humans, plants, and animals, as they are the home of forms of life that endure unique conditions. The plants are physically like those in wetter regions, and contrary to one's expectations they are often extravagant users of water. Their adaptation lies in the ability to survive long periods of want by going dormant. The animals, with few exceptions, show little more tolerance for heat than creatures of colder climates. They survive in the main by evading the hot days with deep burrows or rest, and by living their lives nocturnally. As for man, it seems the people of arid lands feel the pangs of thirst and the furnace blast of heat as much as other humans. The lore of the desert peoples is filled with testaments to the horrors of dehydration and heat prostration.

Modern industrialized communities in such places are oasis cultures living in the desert but not off it. With pumped groundwater, air conditioning, machines and imported food, such centers have been able to remain ignorant of the environment just beyond the city pavement. About 150 years ago, a caravan of 2,000 humans and 1,800 camels died of thirst in the southern Sahara (Cloudsley-

Thompson, J.L., 1965). During the summer of 1905, 35 humans died of thirst in Death Valley alone (McGee, W.J., 1906). When man leaves his modern cities and enters the desert, he is but a few hours from this world and short days from death if he fails to heed its presence. A day without moisture and he will chew anything for water; a few days without food and he will kill anything for sustenance.

"A poor devil on the Mohave desert," W.J. McGee relates in his classic paper on desert thirst, "reached a neglected water hole early in this stage [of dehydration]; creeping over debris in the twilight, he paid no attention to turgid toads and a sodden snake and the seething scum of drowned insects until a soggy, noisome mass turned under his weight and a half fleshed skeleton, still clad in flannel shirt and chaparejos, leered in his face with vacant sockets and fallen jaw; he fled . . ." (ibid.). His trail later showed that he spent the rest of his ordeal seeking again this same spot. When rescued he raved for days "of his folly in passing the 'last water.' " There are many such tales of men drinking their own urine, of Tom Newton who in 1905 went four days without water in Death Valley and was found "aimlessly 'leaping about in the sun like a frog . . . ,' " of a child who wandered into the Mohave and was found impaled on the hundred thorns of a cactus embraced in a terminal surge toward moisture (ibid.).

This is the core of aridity: the danger to biological life. It is a forgotten core, and the mummified caravans, the delirious prospectors, the ruins by the Tigris, the Euphrates, the Nile, the Salt, the Gila, the dead cities of the Iranian plateau, the vanquished communities of the Rajasthan desert—all these scraps of memory and fact are often regarded as a curiosity. The deserts that spawned religions, that drove men to rain songs, that resounded from singing sands with their twang like a monster harp, these places and things are little known or considered by modern systems of exploitation based on mining fossil fuels and fossil water. Blotted from consciousness, their reality seems to many more doubtful than the plenty of desert cities and irrigated farms. Humans have avoided the brunt of aridity by constructing microclimates with steel, electricity, fossil fuels, groundwater, pumps. But this shelter does not promise to be as enduring as the facts of aridity itself. The arid lands are reasserting themselves around the globe, and this process is now called desertification. It describes degradation of land to desertlike conditions because of human activities (Sherbrooke, W., and P. Paylore, 1973). It is a message from the land.

A key problem facing man in the drylands is water, and the problem with the water is that it is largely sight unseen: groundwater. The surface supplies afforded by the few rivers and springs have long been exploited, and future expansion of human activity in the arid lands, indeed much of the current activity, depends on lifting water from the earth. Hydrology is in many ways the history of man's efforts to locate this buried water. Modern techniques and sensing devices have helped in this search, but they have not significantly changed the age-old problem that man looking for water underground largely looks blind. Expensive quests by scientists still sometimes result in dry holes, brackish or saline water.

In recent years a new machine has sought to minimize the hazards of this searching: the computer. Into its maw of printed circuits, massive piles of fact, graph, and notation are poured, then programed and modeled into the closet drama of high technology: simulation. As the pages of printout spew forth, aquifers can be drawn down, wells drilled, water qualities varied, decades of exploitation hypothesized in minutes. With aids like this machine, scholars believe that groundwater analysis has moved from a dowsing rod level to the credibility of a science. One survey of the field concluded that groundwater hydrology "is on the threshold of an important period of integration" (Simpson, E.S., 1967). By converting water, rock, soil, and human consumption to numbers and then rapidly digesting the numbers with the computer it is hoped that a way can be found to deal with the nature of groundwater in arid lands: that it is basically nonrenewable.

More data are sought for the machines on hydrogeologic mapping, the relationship between chemistry and aquifer properties, and the possibilities of artificial recharge (ibid.). Models have been constructed comparing rapid mining of an aquifer with steady-state use (Gisser, M., and A. Mercado, 1972). Water systems for regions can be simulated complete with reservoirs, canals, pumps, rivers, and rains. Under the title "Coping with Population Growth and a Limited Resource," Arizona's chronic overdraft of groundwater has been made a series of optimization and simulation models (Briggs, P., 1972). Sometimes the consequences of various water uses are projected, and, in one case, farmers are sacrificed to urban thirst (Kelso, M.M., W.E. Martin, and L.E. Mack, 1973).

Irrigation districts are modeled, evaluated, planned (Heady, E.O., et al., 1971).

There are reasons for this. Many nations in the arid lands depend on groundwater. Tunisia's water supply is 95 percent subsurface. Morocco must pump 75 percent of its water. Israel is about the same. Saudi Arabia is based almost completely on groundwater. Wells are a major and growing source of water throughout North Africa, Western Australia, the American West, central Asia, India, Iraq, Afghanistan, Pakistan, and other nations (Cantor, L.M., 1967). The use and dependence upon groundwater are growing as population expands beyond the capacity of surface supplies. The United Nations has cautioned that "it is necessary to bear in mind that these natural resources (water and land) are permanent assets of every nation, in fact of humanity at large. These resources have to be used wisely and passed on unimpaired and, as far as possible, undiminished to the generations to follow" (FAO/UNESCO, 1973).

It is hard to fault these efforts to garner more information, simulate various futures, and wisely husband resources lest posterity be deprived of necessities. But when such inquiries are focused on groundwater in arid lands there is a limit to what can be done. If one accepts the hypothesis of those who argue against safe yield as a goal, little need be done since the society which exhausts its groundwater will in this very effort find the economic elements of a solution. On the other hand, any attempt to adhere to the United Nations advice about passing on water and land unimpaired and undiminished requires extremely limited use of groundwater. There simply is not much recharge to aquifers in arid lands because there is not much rain; to use such water sources in the manner typical of Western industrialized societies is to mine them. The information and the machines can help with speculations about how fast to use the water and where to use it. They cannot augment the supply. Or alter the region where it is found.

But the talk of turning the arid lands to account keeps pouring from the presses.[5] Words of caution about the grotesque results of man's hand in such places (Sauer, C.O., 1969) and warnings about the creep of deserts into the croplands (Sherbrooke, W., and P. Paylore, 1973) are drowned by cadenzas of schemes and dreams.

Cheap energy, abundant water, and an army of studies have helped keep consciousness of the arid lands at bay. But this feat has been accomplished at a high cost in resources. As the water tables sink, as the price of a BTU increases, defenses against aridity will gradually decline. For the place has never really changed at all. It has merely been erased in small patches of the earth where humans have concentrated resources. Societies based on the rapid exhaustion of storages face the possibility of lasting just as long as those storages. That is why the place matters. The deserts and semi-arid rangelands are living off of the renewable solar flow of materials. They can wait outside the borders of human cities and farms. Ready to return.

People

The woman remembers. She is fifty something, Papago. Her life has been spent in a village beneath Baboquivari Peak surrounded by kinsmen, desert, mud walls, personal things. Yes, she says, many used to farm, now not many. Her husband is one of the few tribesmen who has clung to the ancient O-otam way of wresting a crop from the Sonoran desert. This way is called akchin: arroyo mouth. It is very simple. In summer, due to tilt of earth and global swirl of air, the rains come to the Baboquivari Valley. Day after day the puff of cloud builds over land electric with the dry. Suddenly, release comes from the bondage of sunny skies: rain. The moisture strikes the land with violence and roars off the rock slopes of the mountains, savaging a route down the arroyos to the valley floor. Here, speed is lost, slope gives way to flat of floodplain and the waters move as a sheet over the porous dirt, sinking in, making life possible. That is when the man plants.

Once the O-otam used a digging stick, then centuries ago a shovel was glimpsed in a European hand and the stick widened. A hundred years ago, Yankees came peddling metal tools. Finally, the iron plow and the horse were brought to the ancient fields. The tilled ground drinks the rain. Then, the man plops his seed in the mud and waits for the sprouting of corn, squash, melon, bean. He hopes for more rain, and the plants race the sun for the moisture in the alluvium. Sometimes he loses and the rains stay away from his part of the valley, and nothing gets planted or everything gets planted and everything dies. Sometimes he wins, and the seed planted in July with the rains becomes a plant guarded in the fall.

The man works but a few acres. In the middle, a hut with mesquite beams, dirt floor, and walls of ocotillo ribs serves his needs. Here he sits in the heat of the day. Here in the fall he spends the night guarding his melons from raiding coyotes. This is the system called akchin: no fertilizer, no pumped groundwater, no insecticide, no herbicide. This way of growing food has all but vanished. In 1914 a federal survey found the Papago cultivating thousands of acres by floodwater farming (Clotts, H.V., 1915). Today, this enterprise is but a memory in the villages. Sometimes the remnants of former fields can be seen; sometimes nothing remains.

Since the 1930s this way of farming and the way of life based upon it have been driven from Papaguería. The old man with his field, his hut, his seeds, his wait for the July rains is a ghost on his own land, and when he dies and the other old men die in the next decade or so, akchin will be gone. It was a way of living in the desert without wells, without pumps, without electricity, without fossil fuels. These very forces have vanquished akchin by making it obsolete and futile in the eyes of the Papago. The young will have none of it and the old have mainly let it go. Here, in the struggle between akchin and the energy systems of the white man during the twenties and thirties, the Covered Wells stick ceases, as if to say that the need for history had passed and the past was no longer usable.

In 1915 the stick says:

> Even while the well rigs were going, copious rains fell making bounteous crops and feed for stock, thus proving the lack of necessity for wells and vindicating the judgement of the wise Indians in opposing the innovation. (Kilcrease, A.T., 1939)

The record continues through the teens, noting the influenza epidemic, ignoring World War I. An aurora borealis lights the skies in 1921 and old Papagos say it is a portent. The rains fail for a year. The following year is marked by renewed faith in akchin:

> The wise men decided that neglect to celebrate the rain feast was the cause of all these calamitous happenings and that it must be done this year. It was done and no more disasters have occurred. (Ibid.)

The stick proceeds for a few years recording items of interest to the Papago. A road is built in 1928; a year later an automobile runs over an Indian. With the thirties, information from the nicks and gouges ends as the Bureau of Indian Affairs and federal work programs finally penetrate this desert world.

Perhaps this too is part of what the woman remembers. She has lived from the world of the stick into the world of today. Her life is a crazy quilt of different energy systems. The house is stucco-covered adobe surrounded by an eight-foot ocotillo fence. The fence is alive and it pulses green or brown with the rains. In summer, life is lived outside under the ramada of mesquite beams, saguaro ribs, dirt-packed roof. Two metal beds catch the night air. A dozen light blue Mexican pots hang; on the ground lies a stone metate for grinding corn. She is making tortillas; the ingredients are kept in two electric refrigerators. The work is done by hand. A metal sheet over a mesquite fire is the stove. The bag of flour on the table is federal surplus. The house nearby has electric lights, dirt floors. Across from the ramada is a chapel such as many Papagos have. Sometimes they contain *santos* centuries old next to crosses fashioned from plastic egg cartons. It is the heat of day and nothing stirs in the desert as the woman works. On the kitchen table is an alarm clock.

The woman knows the desert. On a walk she can show what plants are good to eat and when they should be picked. Like her husband's farming, this information will die with her. It no longer matters. Food comes in packages, energy from electric lines, water from wells.

The Papago today are much like their white neighbors an hour or so to the eastward. There, in the Santa Cruz Valley, three to four hundred thousand Americans live in a modern urban community. It is said to be the largest city in the United States based solely on groundwater. The streets are lined with trees, the yards lush with lawns. People with money build swimming pools. The largest single employer is the government. The inhabitants call their town the Old Pueblo. It is an outpost of money and resources from the east. Food comes from other states. Electric lines reach out almost five hundred miles seeking power. Water heaters run on natural gas from Texas. The homes, schools, stores, and automobiles are air conditioned. The parks have artificial lakes. Local building codes have outlawed mud adobe construction. The Papago lands and the city touch.

This place called Tucson is a triumph of humans over the environment. A child in this city can flick a light switch and help make a powerplant in another state burn 20 to 30 thousand tons of coal per day. Each human in this desert metropolis uses 175 gallons of water every 24 hours (Davidson, E.S., 1973). The community banks a river dry more than 300 days per year. It ignores this fact, like so much else, because it has access to abundant fossil energy from outside the desert. Water is pumped from the ground faster than it is replenished; in fact, the city leads its state in this activity. Each year the water table under the thriving city sinks several feet. Subsidence is expected by 1985 (ibid.). The inhabitants do not keep sticks.

These are two of the desert peoples of the region. Today they live roughly similar lives. They are consumers and what they consume are ancient storages of fuel, water, and soil. The Papago are considered poverty stricken, which means they cannot devour goods at the rate they would like. The Americans are considered affluent, which means their enormous appetites are fed. But both are now aliens to the Sonoran desert itself with its natural flows of resources. Days long past pour from their faucets, feed their families, and power their worlds.

One thing separates them: what the woman remembers. Brief decades ago, almost until the start of the Second World War, the Papago lived in the world of the stick. Food came from wild plants and from the rains, and the rains came from rituals, and the rituals came from a past known only to myth. The O-otam way, now almost gone, dominated the people well into this century. As late as the 1930s one Papago was found innocent of the Gadsden Purchase and believed he still lived in a part of Mexico. The people were left to the land because whites did not want it. Finally, in 1916 a reservation was set up. The Papagos entered into the web of American life having never fought a war against the United States or given ground.

If one wishes to examine the impact of groundwater development the O-otam offer a rare opportunity. Within the lifetime of humans still alive they have moved from a society perfectly adapted to aridity and absolutely independent of groundwater to a society independent of the desert and based on groundwater. For those interested in peace with nature, the Papago offer abundant clues to the price of such a peace. For those fascinated by development, they are an example of how it is done and what it does.

There is a debate going on about development itself. Humans are asking what past achievements mean, and what future plans should be. This debate continues under many guises, such as arguments about steady-state economies, pitches for green revolutions, alarms about the environment, concern over population growth—concern over growth, period. Groundwater leads one into this swirl of contention because it stimulates societies in arid regions it is by nature incapable of sustaining.

The history of the desert O-otam, and the Piman people in general, grants insights into this debate. They have lived its many aspects. It has not been a tidy journey, and the investigation of the move from renewable resources to nonrenewable resources demands a look at the culture of the humans who experienced such a change. To look away from deep wells and coal and oil and electricity is in this instance to peer into a world of dreams, sharing, rituals, akchin.

2. Stick

The Piman people occupied much of the Sonoran desert. Their societies were responses to that desert. To understand them the land that fashioned them must be explored. The desert ranges over 120,000 square miles, about the size of Italy. It is characterized by heat, low rainfall, and high evaporation rates.

Located on the eastern edge of the North Pacific High, the region is locked into a fairly persistent atmospheric stability. Dry, moisture-hungry air rakes the land. Precipitation varies from 3 to 12 inches; evaporation ranges from 6 to 9 feet. This is not the place for lakes and broad sluggish rivers. Rain comes twice a year. Winter storms slip off the Pacific and are slow and extensive. Because of their duration and the time of year, they are extremely important to the desert's water budget. The summer monsoon comes from the Gulf of Mexico in June, July, and August; these rains are often violent thunderstorms. Brief, sometimes covering but a few square miles, the summer storms generate flash floods, high winds, and arroyo cutting. Basically, winter rains decline in importance the farther one gets from the Pacific, and summer rains decline in importance the farther one gets from the Gulf of Mexico. Everywhere rain is uncertain. Yuma, Arizona, averages 3.48 inches per year; one August day in 1909 4 inches fell (Dunbier, R., 1968).

Temperature is more dependable. The summers are hot, the winters mild. The greatest heat and least moisture are found in the western parts of the desert. One spot on the lower Colorado River has registered 136.4° F (ibid.). The land is a patchwork of habitats, temperature regimes, water concentrations. Averages, annual means, and the like obscure as much as they illuminate. Rainfall varies wildly from year to year and place to place. Summer precipitation can be so spotty that a deluge at one weather station may only speak for a few square miles. The mountainous topography of the desert tends to swiftly collect rainfall into arroyos and ephemeral rivers. The fierce sun evaporates any moisture. Periodically summer rains are witnessed which fail to reach the ground. All these factors mean that a one-inch rainfall does not mean one inch everywhere.

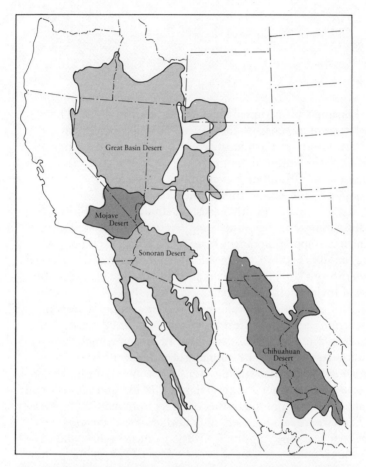

MAP 2. The North American Desert and Its Subdivisions. By permission from *The Changing Mile*, Hastings and Turner, Tucson: University of Arizona Press, copyright © 1965.

It means a lot in a few places and not much for the rest of the
land. Vegetation reflects this reality. Trees choke streams while a
few yards away cacti dominate the landscape. This huge tract of
arid plants is here and there punctured by mountains crowned with
pine and douglas fir.

The topography is described as basin and range, meaning that
it is characterized by roughly parallel mountains (running north-
west/southeast) and intervening valleys. From west to east, the
mountains grow ever higher; they occupy about 50 percent of the
land surface (ibid.). The ranges are a series of walls slicing across
the desert. Surrounded by an ocean of alluvium covered with desert
plants, they exist as islands in an expanse of eroded rock and soil.

The soil has a high salt content. The meager streams of the
desert fail to flush land. Often vertical drainage is impaired by a
hardpan called caliche. Rich in minerals, the topsoil is low in
organic matter—less than 1 percent. Scant vegetation produces
little litter, and violent runoff sweeps much of this residue away
(Shreve, F., 1951). North American deserts are bereft of dense
plant cover. The area occupied varies from as high as 70 percent of
the surface to as little as 8 percent. The exposed soil is the basis
of dust storms, flash floods, high runoff rates, and low levels of
aquifer recharge.

Within these physical conditions flourished the Piman people:
riverine Pima, Papago or desert O-otam, and Sand Papago. They
ranged from the Colorado and Gulf of California on the west to
the Santa Cruz on the east. North they spread to the Gila, and
south they dominated the Ríos Concepción, Altar, and Magdalena
in what is now Mexico. It was an enormous domain; one group,
the Sand Papago, held a portion the size of Connecticut and numbered but two or three hundred.

The edges on the north, east, and south were marked by permanent streams, and these were the lushest zones of the desert. The
rivers were fed by large drainages and snow-crested mountains.
Soil was deep and renewed by flood with silt. Nearby mountains
held forests of oak, pine, and fir. On the desert floor, giant mesquite, ironwood, acacias, and palo verdes thrived on the dependable
water supply. Mountain canyons harbored cottonwood, willow,
ash, and sycamore. Until this century the streams ran year round.
This fostered an abundant wildlife including deer, sheep, antelope,
black bear, wolves, beaver, and possibly bison (Fontana, B.L.,
1974). These fingers of moisture were a major flyway for birds.

The riverine strips resulted from concentrations of soil and moisture; such spots were highly important to animal life in the desert but constituted physically small slices of the total land area. The heart of Papaguería lay away from the rivers. While the Pimas dwelled along the streams, the Papavi or desert O-otam lived south of the Gila, west of the Santa Cruz, and generally north of rivers in Mexico. The eastern border of their villages was the Baboquivari Mountains; peaked by a seven-thousand-foot rock monolith, the range formed a mental focal point for the entire tribe. Past the Baboquivaris the land is classic basin and range topography. The mountains trend north to south and regularly lose altitude as one goes westward: Quinlan, Coyote, Roskruge, Silverbell, Artesa, Comobabi, Santa Rosa, Tat Momoli, Quijotoa, Vekol, Table Top, Mesquite, Sand Tank, Ajo, Batamote Redondo, Sauceda. Finally, the Growlers are reached; here the culture of the O-otam confronted a wall of increasing aridity.

This stretch from the Baboquivaris to the Growlers has numerous valleys and not one living stream. Succulent cacti proliferate, the perennial woody trees (mesquite, ironwood, palo verde, acacia) endure in a gradually stunted form as the rains fall away to the westward. The almost indestructible creosote bush is common. Higher elevations hold agave. This plant cover has earned the Sonoran desert the title arborescent. Dry and hot, the land is somehow cluttered with life. Europeans and Americans that ventured into this botanical wonder often called it desolate. The plants generally sport spines; foliage is commonly vile to the tongue because of oils of varying toxicity. The lushness will starve the ignorant. Here, in concentrations shrinking with the westerly decline of water and vegetation, roamed antelope, deer, desert bighorn, javelina, coyotes, lesser creatures, and man.

This dry passage from towering Baboquivari Peak to the crest of the Growlers was the heart of the desert O-otam world. Most of the known forty-four aboriginal watering places lie in this tract (Hackenberg, R.A., 1964). The desert panorama of succeeding mountains and valleys was the arena for Papago "migrations and wanderings, largely in search of the means of subsistence, of which the first and the second and the third are water, *water*, WATER . . ." (McGee, W.J., 1898). From the Growler Mountains westward this scant water supply of springs and murky holes dropped from meager to murderous.

Excepting a few misguided white ranchers who failed and a few booming mining camps that soon collapsed, no one ever tried to wrest this land from the desert O-otam. The Jesuits, who have left abundant evidence that they could endure almost anything, visited with the cross but did not stay. The nineteenth century Americans, who have left ample record that they would take almost everything, crossed this land on the way to the gold rush of 1849 but did not return with visions of empire and legal papers for homesteads. Modern Americans, who will go almost anywhere in the name of development, have largely stayed away. This place was left to the Papago, and the Papago were left to their fate.

Traveling from the Growlers to the Gulf of California and the Colorado River, conditions grow yet more harsh. The soil is salt loaded because the five, four, or three inches of rain cannot cleanse the earth. Vegetation similar to that common in the central part of Papaguería is rarer and stunted when found. More than 90 percent of the plant cover is creosote or bur sage. Desert sheep persist, but they can go a week without water and wander forty-odd miles in search of it (Fontana, B.L., 1974). Other animals exist in diminished numbers compared to populations to the east. Eagles nest here, wolves and mountain lions periodically penetrate. By the ocean, life thrives in such forms as sea birds, fish, turtles, sea lions, clams.

Water sources of this region can almost be ticked off on the fingers of the hand. At Sonoita, Sonora, a clear stream one foot deep and about twelve wide runs a short distance and then melts into the sand. Below this disappearance, moisture can be dug from the dry bed of Agua Salada, Agua Dulce, and El Carrizal. The rare rains fill Cristobal Wash and Growler Wash for a spell. A pond 150 meters wide exists at Susuta, Sonora; in the dune country to the west at Laguna Prieta a similar puddle can be found. Las Playas, a dry lake west of Sonoita and the Growlers, is briefly wet after a shower. A few springs trickle from the mountains. The main hope for water is the rock tanks. These storage vessels in the eroded stone catch and hold the rain for a few weeks or months at a time. Nine lie in the Tinajas Altas Mountains, five in the Cabeza Prietas, a couple in the Sierra Pinacates, and one each in the Tules, Pintas, and Growlers. Capacity ranges up to a few hundred quarts. Finally, there are the kiss tanks—small rock depressions in shady spots where dew collects (Ives, R.L., 1962). Such are the water resources for an area larger than some states (Fontana, B.L., 1974).

With the gold rush of 1849 this list of springs, tanks, and dew points stabbed into the American consciousness. Humans going west to the bonanza cut across this tract in the hope of saving miles and days. The route became known as El Camino Diablo, the road of the devil. Anywhere from a few hundred to a few thousand died of thirst (Hackenberg, R.A., 1964). A stretch of the road thirty miles long once held sixty-five visible graves; a traveler in 1925 counted fifty burial sites at the foot of the Tinajas Altas Mountains (ibid.). A single rock pile harbored the bones of an entire family; nearby lay the remains of their dead horse and broken water bottles. Here, at the turn of this century, W.J. Mc-Gee, traveler, student of things, ventured. A man, Mexican and trained up in the region, got lost with a one-day supply of water. He was lost eight days. Resting at dusk McGee was aroused from his sleep by a strange piercing rumble. It reminded him of a bull. The man was found nearby, skeletal, hair suddenly gray, features warped by cracked, swollen tissue. This was the root of McGee's classic paper on desert thirst.

Such was the land that molded the Papago. Here they learned to almost worship the fruits of succulent cactus, tap lizards on the head with ocotillo wands, and savor water moderns find beyond the range of human thirst. From this ground emerged the water-obsessed tongue of the O-otam. "That's the way First Born prepared the earth for us," their legends explain. "Then he went away" (Saxton, D., and L. Saxton, 1973). Or as the U.S. Geological Survey put it in a 1925 guide to desert watering places:

> The so-called Papago country, lying south of Gila River, between Tucson and Yuma, has had until recently only a few white inhabitants and has seemed so waterless and formidable a region that it has been rarely visited by white men. It has been a sort of strange wonderland, isolated and different from the rest of the country—a desert that has perhaps taken a larger toll of human life than any other arid section of the United States, yet green and tree-covered and for unknown generations providing a home and livelihood for a simple-hearted, peace-loving tribe of Indians. At one time the entire region was sparsely inhabited by the Papago. . . . (Bryan, K., 1925)

People

The Papago were a Piman people, and the Piman people stretched across the land south of the Gila River for a thousand miles (Spicer, E.H., 1962). Broken into various groupings (Upper Pima, Sobaipuri, Lower Pima, Tepehaun), their worlds were separate, their languages mutually intelligible. The aboriginal population of what is roughly the Sonoran desert has been given at 155,000 (Sauer, C.O., 1935), 150,000 (Hastings, J.R., and R.M. Turner, 1965), and 120,000 (Spicer, E.H., 1962). Pimería Alta, that part of the desert from Magdalena, Sonora, to the Gila and west to the Gulf, held maybe 20,000 riverine people, and 10,000 desert Papago. In all, perhaps 30,000 humans spread out over 60,000 square miles: a half a man a mile (Sauer, C.O., 1935).

This world without a groundwater problem drove people to four basic responses. Those lucky enough to dwell in spots of constant water and steady food supplies lived in villages and pueblos. The Hopi and Rio Grande groups typify this possibility for the southwest as a whole. In Pimería Alta, the riverine Pima on the Gila approximated it. Without dependable living streams, humans tended toward floodwater farming, summer and winter villages, and a mixed diet of cultivated and gathered food. Such groups were called the rancheria people. This was the way followed by the bulk of the O-otam. A third option was the life of the band exemplified by the Apache. This strategy entailed limited farming and almost endless motion in quest of food. Strip away the slight efforts at cultivation and the final tribal choice emerges: the nonagricultural band. In the entire southwest, only a few thousand humans at best attempted this kind of existence, where nothing was grown and everything was clubbed, foraged, or stolen. To face the natural rhythms of the Sonoran desert independent of crops was to risk death. The Seris of Sonora were the masters of this life.

For those who wonder what the limited vegetation, erratic rainfall, high evaporation rate, and almost absent surface water mean for humans, the Seri afford clues. Among all the groups of the Sonoran desert, they made perhaps the least effort to escape its wildly oscillating conditions. For the Seri, life pretty much began anew each day without larders or wells to soften the challenge.

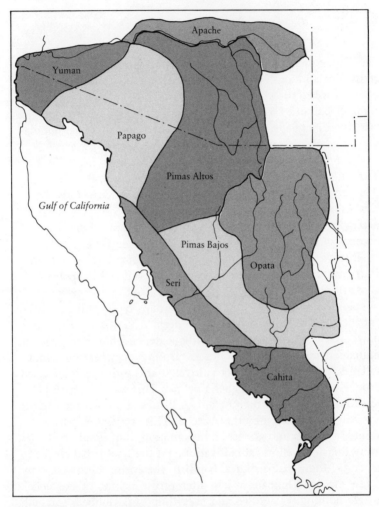

MAP 3. Aboriginal Peoples of the Sonoran Desert. By permission from *The Changing Mile*, Hastings and Turner, Tucson: University of Arizona Press, copyright © 1965.

In the entire southwest, only a few thousand humans at best attempted this kind of existence where nothing was grown and everything was clubbed, foraged or stolen. The Seris were the masters of this life.

Candelaria, a Seri, models her pelican-skin robe. Smithsonian Institution Photo No. 4265-A-1.

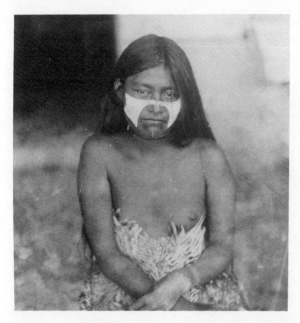

Ampharita, sixteen, Seri. November 3, 1894. The picture cost one harmonica. Smithsonian Institution Photo No. 4273.

Minutes later a Seri meal of raw pork. Found dead from disease. Smithsonian Institution Photo No. 4276-B-10.

Papago home, 9 ft. in diameter and 5 ft. high. Smithsonian Institution Photo No. 2779-P.

A Seri reports on his manhunt for a Seri: "Here I am, Soce Paolina de la Cruz. El Colusio Fernando told me to kill Matalo el Castro, because the informer Ramón López was killed, and it was also Castro who killed him. If we kill Castro, things will be fine; also, if we kill Castro, there will be no more robbers, except the female head, Juana Marinaselle. This is what I told the men, and I want them to lend me a horse." Smithsonian Institution Photo No. 4279-A.

Fifty-five Seris and one ethnologist, W. J. McGee. Smithsonian
Institution Photo No. 4275.

Seri family. Smithsonian Institution Photo No. 4276-A-2.

Seri man, bow, house. Smithsonian Institution Photo No. 4263.
Seri "runner." Smithsonian Institution Photo No. 4264.

Since they grew nothing, they knew the value of most wild plants. They used ninety-two separate species for medicinal purposes alone (Felger, R.S., 1975). Always on the move, they made probably the lightest ollas in the Southwest (McGee, W.J., 1971). Safety dictated that they camp anywhere from three to fifteen miles from water, and simple fetching took a good share of their time. The list of the hydrologic resources is brief:

> The aggregate daily quantity available during ten months of the average year (excluding . . . the two moist seasons) . . . 60,000 to 125,000 gallons per day, of living water, i.e., less than the mean supply for each thousand residents of a modern city [in the 1890's] . . . Probably two thirds of this meager supply is confined to a single rivulet . . . in the heart of Tiburon [a Gulf island] far from the food-yielding coasts, while the remainder is distributed over the 1,500 square miles of Seriland in a few widely separated aguajes, of which only two or three can be considered permanent; and this normal supply is supplemented by the brackish seepage in storm-cut runnels . . . This scanty aggregate serves not only for humans but for the bestial residents of the Seri principality, and its distribution is such that the mean distance to the nearest aguaje throughout the entire region is eight or ten miles, while the extreme distances are thrice greater. (Ibid.)

Limited water meant limited food, and Seri appetites have become legendary. W.J. McGee on a visit in the 1890s saw a child rip a pup from a dog in the act of birth; he devoured it like "a peeled banana." The Seri's sole mode of food storage was the "second harvest." They saved their feces so that undigested seeds could be reconsumed (ibid.).

So far as can be learned, the Seri lived in balance with their resources: zero population growth, zero economic growth, a steady-state society. They never numbered more than a thousand or so, and their population size in relation to their theoretical resources makes graphic what such much-used terms as *carrying capacity* can mean. McGee, during his late-nineteenth-century visit, found that the 300 adult Seris devoured in the neighborhood of 27,000 pounds of cactus fruit per year (saguaro, cardoon, prickly pear). This weight converts to about 216,000 fruits, or 90 pounds and 720 fruits per adult. Saguaro cactus averages

60 fruits per plant, and about 6 plants per acre. Assuming the
Seri harvested only saguaro (and they did not), the tribe would
need only 600 acres of such vegetation. This is less than one
square mile. Projecting this consumption to the Sonoran desert as
a whole, 100,000 adults (undoubtedly a high figure) would harvest only 400 square miles of saguaro, an area 20 miles by 20
miles—this in a desert of 120,000 square miles (ibid.; Hastings,
J.R., and R.M. Turner, 1965).

The skimpy exploitation of succulent cactus by the Seri is typical of human actions in societies lacking access to such storages
as groundwater and fossil fuels. First, the natural concentration
of resources is erratic, and in the arid lands highly erratic. To base
a population on the bounty of good years full of rain promises
death in other less fortunate periods. Secondly, cactus not eaten,
water not drunk, trees not cut, and so forth, are not actually lost to
humans. These resources nourish other parts of the natural system,
and humans are dependent upon the health of the system as a
whole.

It is a moot point whether the Seri and other aborigines realized
this fact; they lived it. But it is precisely this fact which has frequently eluded and baffled Europeans and Americans. They observe
slash-and-burn agriculture and wonder why most of the forest is
left alone most of the time. In the case of the British in Burma, the
apparently unused forests were actually seized and timbered since it
seemed it could not matter to the natives. It did matter. The land
which appeared wasted was simply lying fallow. There are numerous such instances (Boserup, E., 1965). Of course fallow can be
abandoned when humans apply fertilizer, just as pastures can be
shrunk when humans substitute tractors for horses. But without
access to storages, such as petroleum for fertilizer and tractors, consumption for man must remain low lest the flow of food, water,
and fuel become too slight because of radical interference or lethal
because of fluctuations resulting from weather.

The small human population of the pre-European Sonoran
desert is something hardly realized by modern residents. The city
of Tucson, Arizona, certainly contains more people than the aboriginal desert did, probably more than the entire pre-Columbian
southwest. This feat is possible because the current inhabitants live
off nonrenewable resources and the original inhabitants lived off
renewable resources.[6]

While the Upper Pimas (riverine Pima, two-village rancheria

Papago, and bandlike Sand Papago) lived in rough balance with the resources of their land, they practiced different responses as their domain grew more arid south of the Gila and west of the Santa Cruz. Past the Growler Mountains in the country of the Sand Papago it approached Seri-like hardships. Here the rains were so slight, the tools and structures of man so futile, that humans could only survive by rolling with the rhythms of the year. In this immense piece of land lived a few hundred people who roamed searching for food in groups of eighty or ninety. For centuries they succeeded.

Their tracks are everywhere. Paths can still be seen where humans trekked to the salt deposits of Adair Bay on the Gulf (MacDougal, D.T., 1908). Near a tank, the rock way is worn a foot deep. The volcanic Pinacates are laced with routes; signs of seed grinding and camps are discernible. But there are no ruined buildings or abandoned altars. Aside from one small patch of ground annually gardened in the Pinacates, the Sand Papago did not till the earth. They were nomads.

Their constant coming and going reflects the area's hydrology and weather. No modern has ever chosen to disagree with the Sand Papago assessment. Structures identified as marks of civilization elsewhere here spell death. Build a house and it is a trial to survive while constructing it, lethal to linger afterwards. Dwell in fellowship with large numbers of other human beings and mass starvation greets the effort. Become enamored of possessions, and the simple transport of property guarantees a swift end to desert living. Humans in such an environment manage to leave little behind but footprints. The Sand Papago took with them to extinction their major legacy: information.

They ate what they could find and overcome. Probably, the mainstays were seafoods, reptiles, insects, jackrabbits, and the like (Fontana, B.L., 1974). During high tides, they waited in the rising waters for fish. Clams were devoured, and when low tides permitted, oyster beds sacked. They built no boats or rafts. Now and again sheep, deer, and antelope were slain. Birds were trapped. Sheep were killed with patience, deer and antelope simply outwalked. Native plants, especially the sand root (*Ammobroma sonorae*), were the core of their diet.

Clothing was made from hides, and hides were tanned with the root of the torote tree. Sea lions died so their feet would have sandals. Baskets were woven from torote, willow, and bulrushes.

Material for bows came from the banks of the Colorado River,
arrows from arrow weed (*Pluchea sericea*). Badger hair was
twined for string. Home was at most a low stone corral to break
the wind. Often, the people simply sprawled on the earth (ibid.;
Hackenberg, R.A., 1964).

Early Europeans felt that the Sand Papago were exceptionally
barbaric. When the great Jesuit missionary Father Kino stumbled
upon them in the late seventeenth century, his reception, by a group
of women found foraging, expressed the distance between the two
groups. The women were wearing brief aprons of rabbit skin; they
pulled them off and ran toward the priest offering them as gifts.
The Spaniards wondered how humans could survive in such a place
"full of rocks and all kinds of brush and cacti—an arid sterile land
with no water for pasture" (Manje, J.M., 1954).

Today the answer is practically beyond reach. Shreds of evidence
remain. The Sand Papago cultivated a small field in Pinacates; they
presumably ate plants and animals. When Europeans, Mexicans,
and Americans ventured within their domain, they robbed and
killed them. In turn, these strangers slaughtered Sand Papago and
exposed them to new diseases. A traveler in the region in the early
twentieth century found them remembered as "rapacious and probably merciless to strangers, whether Indian or Mexican" (Fontana,
B.L., 1974).

By that time they were just about extinct. One last survivor was
a hermit, Carvajeles. He lived in a cave near Tinaja de los Papagos,
at first with a woman, then alone. He grew a few beans and squash
in the ancient field of the Pinacates and grubbed for sand roots in
the dunes. Once a year he traveled to Sonoita, Sonora, for a little
flour, sugar, and tobacco, and for an annual drunk. Around 1912
he slipped from view; since then the Pinacate country has been
empty (Ives, R.L., 1964).

While they held sway in the desert west of the Growler Mountains, the Sand Papago exported a few items from their lands. Each
year they would walk to the Colorado River and sell the Yuman
people salt, seashells, and their minds—for their ultimate resource
was their experiences in surviving this severe stretch of the Sonoran
desert. From this material, they made songs and dances prized by
others. Those living by permanent water bought messages from
aridity. In a world so bereft of water only ideas traveled well.

This legacy is now lost. Modern societies, growing jittery over their dependence on fossil fuels and fossil water, are becoming increasingly interested in the kind of information that was standard equipment for Sand Papagos. This group knew how to exploit the renewable resources of the Sonoran desert, plants, and animals. Many of the questions now popping up in the field of ethnobotany were answered by Sand Papago efforts to stay alive. With the destruction of this group the last incident of true nomadism perished from the North American continent (Fontana, B.L., 1974).

Now their land is uninhabited. Part of it is a national monument because a type of succulent cactus rare to the United States grows there; part is a game refuge because no one can figure out how to exploit it. Much of their ancient fiefdom is a target for the American Air Force—a place of littered brass and detonations. The humans and their knowledge are long gone.

East of the Growler Mountains and away from the rivers flanking Papaguería lived the two-village rancheria Papago. There were more of them than the Sand Papago because their terrain had more rainfall, and hence more food. Estimates range as high as 10,000 for their numbers, but probably there were never more than 4,000. These people were not nomads. They migrated between summer and winter villages because the desert's level of productivity and patterns of water concentration made this a necessity. A basically sedentary people, they were driven to periodic moves by an environment which precluded fixed residences.

The sedentary drives evident in the two-village Papago were fulfilled in their kinsmen, the riverine Pima. Living in fixed villages along the Gila, Santa Cruz, San Pedro, San Miguel, Magdalena, Concepción, and Altar rivers, this group enjoyed permanent housing and fixed fields. They tilled the floodplains, practiced irrigation from the dependable streams, and dwelled in villages of from 20 to 50 houses, 100 to 600 inhabitants (Fontana, B.L., 1974). As many as 800 humans may have lived at San Xavier del Bac near the site of modern Tucson. These Pimas were comparatively rich in property, and, since the streams were natural corridors of trade, had access to turquoise and shells. "The character of these Indians . . . is haughty and proud," according to the Spaniards, "a fact which can be recognized in the manner in which they talk— with little esteem—about those Indians of the west. These western Indians . . . consider them superior and look up to them with special respect" (Mange, J.M., 1954).

Such were the residents of Pimería Alta. They were united by language and culture, divided by gradations of moisture. From Adair Bay to the Gila a common tongue was spoken and techniques of exploiting the desert held pretty much in common by all. What separated the nomadic Sand Papago from the haughty Gila Pima was resource concentrations, particularly water. They were in a sense one people acting the different possible responses to aridity. Along with the Seri to the south they provide a clue into the costs of consumption balanced with resources. The calendar of the Papago demonstrates this cost. May was called the "painful moon" because the food produced by winter precipitation was gone, and the downpours of the summer monsoon had not yet arrived. The new year began with July and the rains. Harvest came in October and November. January, the launching of the white man's cycle, slipped through the Papago calendar with a simple notation: "animals thin." February was smelling in recognition of the rut of the deer. Then green moon in March recorded the leafing of the mesquite. The flowers of April were noted as yellow moon. Then, May, hunger, and the return of "painful moon." June was called black seeds after the saguaro fruit. By making wine from this bounty, the Papago made magic and beckoned the July rains and a new year (Underhill, R.M., 1939).

Their calendar reads like a menu. Humans dependent on the Sonoran desert for food and drink have just cause for drifting off into obsessions. Every O-otam knew in his bones the region's ratio of producers to consumers, of leaf to deer and deer to man. Scholars now struggle to measure desert rates of productivity. The people of Pimería Alta kept track of such rates with their stomachs. Their art was dry, their architecture dry, their survival dependent on relentless sun and erratic rain.

Food

Two things are known about the diet of inhabitants of the Sonoran desert at the time of the European entry. People ate almost anything they could lay hands on, and their pursuit of nutrition seemed to have little if any impact on animal and plant communi-

ties exploited.[7] One eighteenth century Jesuit, Father Ignaz
Pfefferkorn, wrote a two-volume work on Sonora replete with
horror at what the natives considered good fare. "Often," he recalled, "I have observed with astonishment an animal feast among
the wild Papagos." In times of hunger they would come from the
desert to his mission, "but as soon as they learned there was a dead
horse, burro, or other animal anywhere, they hurried to the spot,
tore it up like hungry wolves and ate it greedily, intestines and
all." The priest was appalled that his charges were "little concerned
with whether the meat is fresh or half rotten, stinking and full of
maggots . . ." (Pfefferkorn, I., 1949).

Then there were the rats. Brushy spots were infested with
rodents, and sometimes twenty or thirty Papago would surround
such a rat empire, set it ablaze, and club the rodents as they fled or
gathered their roasted carcasses if they moved too slowly. "An
Indian wishing to make me a present once offered me some
pieces of this booty," Pfefferkorn remembered. "When I begged
to be excused because I was not used to such delicacies, he wondered that I was so ignorant of dainties" (ibid.). This clash of
palates represented two different water budgets. Pfefferkorn lived
in the Sonoran desert, but he had access to the resource flows of
Mexico as a whole, and of Europe. The Papago, with his temporary
bonanza of rats, ate what the desert offered, or not at all. With the
exception of certain taboos (varying from group to group), the
aborigines were willing to entertain any form of organic life as a
source of fat, sugar, carbohydrate, protein.[8]

These natural flows of energy are today largely unknown and
ignored. Moderns content themselves with purveying a few cactus
jellies and propelling bits of lead through the bodies of random
deer. The O-otam lived the net yield of the desert's hydrologic
budget; their farming and foraging holds a wealth of information
about surviving in the Sonoran desert without subsidy—whether
from the viceroy in Mexico City or from a natural gas well in
Texas. Part of the impact of groundwater development lies in the
contrast of this world of the digging stick with the current world
of the tractor.

"Agriculture," according to Pfefferkorn, "is pursued by the
Sonorans in a very perfunctory fashion." By the eighteenth century plows had reached some Indian hands, but the priest lamented
that "no one knows how to plow a regular furrow. It matters little

to them whether it be straight or crooked" (Pfefferkorn, I., 1949).
Also, the natives failed to grow more than was necessary "for their
yearly maintenance" (ibid.). Indian agriculture often appeared
piddling to Europeans because it sought to adapt to the conditions
of the Sonoran desert while successive waves of Spaniards, Mexicans, and Americans struggled to overcome or escape those conditions.[9]

Early agriculture in the Southwest had one striking feature: a
very short growing season. One expert concluded that "this area of
climatically long growing season was for pre-irrigation agricul-
turists the shortest season agricultural area in America" (Carter,
G.F., 1945). The long frost-free period so attractive to modern
white farmers held few joys for aboriginal tillers of the soil. Crops
could not begin until the rains came, or the rivers rose. The Hopi
of the Colorado Plateau illustrate this point. Technically, the average date on their mesas for the last killing frost is May 19, and this
gives them 133 days until the next frost. But the rains do not come
until July; the Hopi actually have only 90 days between rain and
frost. Most of their crops require 120 days for maturation. These
are the real nightmare parameters of Hopi farming.

The Cocopa, Mohave, and Yuma people of the lower Colorado
River faced the same problem. Living in one of the hottest parts of
North America, the frost-free period along their stretch of the
river ranged from 251 days to 357 days. But the Indians grew only
a single crop per year. Rainfall, less than five inches per annum,
was hopelessly inadequate. Farming depended upon the Colorado's
yearly flood. This blessing came in May or June, but it was usually
early July before the receding waters permitted planting. Seeds
were buried in the mud, and then a race began. The plants had to
produce before the fierce sun dried out the mud. This left the Colorado tribes with a 60-day season (ibid.; Castetter, E.F., and
W.H. Bell, 1951). Their crops of corn, squash, and tepary beans
were characterized by rapid growth. Experiments have shown, for
example, that teparies germinate in half the time of kidney beans;
modern corn, puebloan corn, and pinto beans fail under the conditions that the Colorado tribes faced (Carter, G.F., 1945).

Agriculture among the riverine Pima and two-village Papago
faced the same problems. The O-otam spent half the growing season waiting for the July rains; the Pima on the Gila, Santa Cruz,
and other rivers were hostages of the flood. Records of the pre-
dammed Gila suggest what this means:

January	50,000 acre feet	July	24,000
February	52,000	August	39,000
March	56,000	September	26,000
April	36,000	October	24,000
May	12,000	November	17,000
June	3,000	December	34,000

(Castetter, E.F., and W.H. Bell, 1942)

About every fifth year the Gila failed completely (Russell, F., 1908). Sometimes the floods were too mighty and crops were lost, or replanting was a necessity. As with the Colorado tribes, the timing of the floods sharply reduced the Pimas' hypothetical 263-day frost-free growing season (ibid.). Until the Spaniards arrived with wheat and barley, the Pima lacked a winter crop capable of utilizing the heavy flows of that season. Such were the terrors of living with the desert's rhythms.

Whenever possible the riverine Pima and desert Papago practiced canal irrigation. The Pima lined tens of miles of stream bank with fields, ditches, and crude dams. The dams, made from mesquite poles and rocks, rarely survived more than a year. Ditches off the stream might run for a mile or more and were often ten feet deep and four to six feet wide where they left the river. Fields were leveled for even watering and broken by ridges into plots one or two hundred feet square. All this earth was dug without shovels, and moved by baskets. The Pima fields were handmade solar collectors.

Men were organized into work groups (ibid.). "All this work was done under a supervisor—sometimes they needed two for a big job," according to the twentieth century Pima. "When they got through at night the boss would lecture them on how good work they had done. He would tell them that now they should go home and plant more than last year and try to make good. He encouraged them. Everytime they work together on a job he makes a speech when it is over" (Hackenberg, R.A., 1964). A seventeen-mile ditch on the Papago lands proves that they made similar efforts. But the undependable stream flow seems to have driven them to thunderstorm farming (Castetter, E.F., and W.H. Bell, 1942).

Farming the rains required from seven to ten inches of annual precipitation (Hackenberg, R.A., 1964; Carter, G.F., 1945; Castetter, E.F., and W.H. Bell, 1942). Of course spacing was as

crucial as amount. Excepting gourds grown periodically for utensils, neither riverine Pima nor desert Papago hand-carried water to
their plants. Both groups lacked rituals tied to irrigation. One
modern informant explained by saying farming was not the place
for ceremonies but for brainwork (Castetter, E.F., and W.H. Bell,
1942). In the same vein, a twentieth century Papago decried abandoning eroded land; he said a man should stay with his field because there was nothing to be gained by moving elsewhere (ibid.).

The fields, whether Pima or Papago, varied in shape and were
fenced with stakes (McGee, W.J., 1898). Tilled with a digging
stick, the rows of hills showed no particular zest for straight lines.
If a tree stump was encountered, the plants staggered around it.
Only the hill itself was worked. Randomness changed the location
of the hills from year to year and helped extend soil fertility. Seed
depth varied with the crop, and four being a ritual number among
the Pimans, seeds were planted in multiples of it (Castetter, E.F.,
and W.H. Bell, 1942).

The work was brutal. Because the crusty soil often inhibited
seedlings, the Indian had to break it up, hill by hill. Insects and
other pests were stalked by humans on a one-to-one basis. A man
rose at dawn, went to his fields, and stayed until dusk. During the
heat of the day he rested under a mesquite tree. Cultivation was
endless. All this toil netted yields almost pitiful by modern standards. Indian corn returned about ten to twelve bushels per acre;
tepary beans produced five hundred to eight hundred pounds per
acre. This was the reward for humans who spent months crawling
and chopping on their hands and knees (ibid.). Cropland per
family averages from a quarter acre to two acres depending upon
the richness of the site. Papagos got perhaps 25 percent of their
food from agriculture, Pimas half or more (ibid.).

This garden was not Eden. The ingredients of successful agriculture had to be delivered by man or nature. Enormous tracts of
desert concentrated the thin rains on tiny plots for brief spans of
time. Fertilizer meant flood-delivered silt. Insecticide meant a man
reaching out with a hand to crush a bug. Herbicide was the swing
of stick. Ditch, dam, canal, cultivation, planting, harvest—all were
reduced to muscles in the back, the arms, the legs. The oscillations
of the desert year could not be leveled with pump, tractor, spray,
and icebox. The man with the stick could not reach the aquifers,
the coal beds, the oil wells, the gas reserves. He had no groundwater problem. Life in this solar-fired world meant resources were

renewable and the delivery of resources erratic. Humans reaped positive yield for their labor and endured periodic want as a reward. Their inability to mine nonrenewable resources limited their agricultural production and kept them dependent upon the yield of undomesticated desert plants and animals. To eat they must forage.

The problems of the field persisted in the desert terrain. Annuals flourished or grew not at all, depending upon rainfall. The mountains and drainages that concentrated water for agriculture had the same effect upon uncultivated plants. The yield of the various shrubs fluctuated across the land because water was not uniformly distributed and could soar or collapse from year to year with the precipitation. Some perennials with deep roots, such as mesquite, and some succulent cacti with massive water storage, such as saguaro, evened out these oscillations to a degree. The Sonoran desert, like the tilled plot, was a productive gamble; safety for the aborigines lay in the desert's sheer size. The odds were improved because there were chances for the necessary combination of sunlight and water.

The natives of the Sonoran desert exploited at least 225 different species of wild plants, with about 15 percent being major food sources. The ephemerals were rain dependent but had a tendency to grow swiftly and put a high proportion of their energy into seed production. "In other words," it has been argued, "the ratio of seed to vegetative productivity is unusually high" (Felger, R.S., 1975). Probably, the diverse sources of nutrition ensured a healthier diet for the Indians than would have been possible with agricultural monocultures (Ucko, P.J., and G.W. Dimbleby, eds., 1969). The escape from the perils of the garden through food gathering gave the Piman people a striking feature: motion. They pursued the vegetative pulses across the face of the land.

When the new year began in June, the people of Papaguería moved from their winter villages by permanent springs to the cactus camps down below. The saguaro (*Cereus giganteus*), by drawing on stored moisture in its tissue, offered forth fruit weeks before the summer monsoons began. Eaten fresh the fruits were sweet, boiled they made a syrup, fermented they offered a wine, cooked down they would make up a jam. Oil could be pressed from the seeds or the seeds could be ground into a flour. The cactus' wooden ribs wound up as racks, screens, doors, bird cages. Scar tissue boots caused by bird damage sufficed for vessels. Spines became needles for tattooing (Hackenberg, R.A., 1964).

Following the orgy of the saguaro harvest, the rains came and crops were planted. This done, the Indians turned to other wild plants. July was full of lambsquarter (*Chenopodium murale*), the stuff of a vegetable stew. Greens were garnered from pigweed (*Amaranthus*), an annual. Come September this same plant produced seeds extremely high in protein and oil (Felger, R.S., 1975). Bur sage (*Franseria*) also was eaten as a green; in some types the roots and stalks were harvested. Datil (*Yucca baccata*) offered an important fruit; the leaves were split for basket material. Seasoning was found in the branches of saltbush (*Atriplex*) and the fruit of bird redpepper (*Capsicum typicum*). Various chollas and prickly pears (*Opuntia echinocarpa, O. engelmannii, O. fulgida, O. versicolor*) yielded buds and fruits the Indians pit baked. In the case of the prickly pear the pads were also devoured. During July, parties ventured into the higher elevations for acorns (*Quercus emoryi, Q. oblongifolia*). Organpipe cactus, which bore fruit twice yearly, was also exploited during the summer months.

With August the beans of the perennial desert trees became available. Paloverde and mesquite were a major food source for the Piman people. The trees were numerous, hardy, and if favorably located for water, prolific. Their woods (as well as ironwood) made excellent fuel and a hard material for utensils. Though never cultivated, favorite mesquite trees would be returned to annually because of unusual fecundity or a unique flavor (Felger, R.S., 1975). Mesquite bean production fluctuates a good deal with the rains; sometimes the crop can almost fail (Parker, K.W., and S.C. Martin, 1952). The seed is highly nutritious. *Prosopis juliflora*, the kind common to southern Arizona, has a bean protein content of from 34 to 39 percent. This is equivalent to soybeans. Used as a flour, the mesquite was the staple of the Sonoran desert. Harvests of paloverde, catclaw, and ironwood supplemented this key resource (Felger, R.S., 1975).

Fall brought the seeds of grasses; in addition, prickly pear continued, and in the case of the Sand Papago, comotes or sand root were dug. By October water scarcity began to drive the desert Papago to their winter, or well villages. Here they faced the cooler months with the harvests from their cultivated plots, the several wild food sources. Agave or mescal entered the diet in December. The crown of the plant was gouged out, then roasted. Leaves went for fiber and soap. By March, several more eatable plants were ready. Wild rhubarb (*Rumex hymenosepalus*) made pies and

offered a cold remedy. Papago bluebell (*Brodiaea capitata*) provided an onionlike bulb. Considered distasteful by the desert
people, it was consumed because food was scarce at that time of
year. Basically the entire winter was a contest between Indian
storages from the summer crops and the wait for the next monsoon. Usually, the Indians lost, and May, month of the painful
moon, meant sufficing with wild onions, cholla buds, and sotol.
Then came the flowering and fruiting of the giant cactus and the
world began again.

Many other plants were exploited for special purposes. Datura
or jimsonweed functioned as a medicine and perhaps as a mind-
altering drug. Arrowweed gave arrows. *Coursetia glandulosa* became a gum for sealing jars of saguaro syrup. And so forth. This
search for food, fiber, and medicine among the flora of the Sonoran desert was essential for enduring the region's hydrologic budget. Summer rains were erratic and short term. Winter rains were
gentle but came at a time inconvenient for agriculture. Most of the
year was dry skies. Desert plants were a bridge between these two
seasons of rain.

So were animals. The rats, pigs, deer, sheep, lizards, insects,
rabbits, and others gave the Indians access to plants they could not
eat and stored energy they could not preserve. The mesquite not
harvested, the saguaro fruit not picked, came back into Piman
hands as meat and fat. Sheep wandering a week without water, rats
combing the litter of the earth's surface, deer feeding off the steep
slopes shared these traits with humans when they were killed.
Through their flesh, man consumed things he could not reach on
his own and digested things his stomach could not tolerate unaided.
It was much the same with the plants. Lacking pumps, the Indians
used mesquite with roots penetrating the earth more than one
hundred feet. Incapable of building water towers, humans depended upon the gigantic saguaro with more than 90 percent of
their bulk moisture. The wild swings of the year were in part
evened by the plants and animals, and dilute resources were concentrated.

The difference between this dead world and the one now surging
across the Sonoran desert lies in the amount of energy and water
expended in achieving the same ends. The Indians, filling their
bellies with rats, mesquite beans, and flower seeds, were trapped
into exploiting renewable resources. When humans manage to get
access to concentrated resources (oil, water, metal, etc.), they

escape the painful, natural variations of annual concentrations. Often, the problems inherent in this exploitation of storages (imbalance with rates of replenishment and pollution) are dismissed because the humans feasting on the temporary plenty believe that valuable parts of their society are based on such use. Democracy, widespread education, mobility, opportunities for changing careers, and chances to raise personal income are commonly described as dependent upon current levels of consumption and upon economic growth. It is said that the water table must decline so that the possibilities of human life may expand. This argument can be projected back into the past. The way of O-otam was based on their use of resources, and it was as different from the nature of contemporary society as their diet. The humans without wells conceived a world all but unthinkable to their successors in the Sonoran desert.

Way

Because nature was not regular, the ways of man had to be. The desert of change, surprise, uncertainty, produced very steady humans. The girl would be thirteen, fourteen, fifteen, her body becoming that of a woman. The mother and father would notice her breasts getting large. They would say, "They should be used for something" (Underhill, R.M., 1939). It was time for marriage. Piman youth reached adulthood without the experience of adolescence. Entry into the work force could not be postponed for anything like the teenage years characteristic of modern industrial societies. Privation was too near. Old Papagos in the 1930s still recalled times of such hunger "that a person became dizzy when he stood up" (Castetter, E.F., and W.H. Bell, 1942).

Until the age of ten the child was left a child. "We went to a sunny place," one old woman remembered, "and played husbands and wives with our dolls. We girls would make our dolls grind corn. The boys would take their dolls and say, 'We go hunting' " (Underhill, R.M., 1939). Children talked when adults were talking, ran among the dancers in a ceremony, shouted during holy songs. "We don't mind these things," Papagos explained during the 1930s; "we know the children don't understand." Childhood ended with work. Girls toiled with their mothers, grew silent or disappeared if strange men were present. Boys ran errands for miles, and learned to be silent around adults.

A Gila Pima recalled it this way: "The old people told them not to steal, not to go out nights, get up early and go to work, always kill Apaches and help everyone who needs anything" (Hackenberg, R.A., 1964). Lessons came as if from a dream. Night has fallen and the family is lying in darkness on their mats; sometimes it is the hour just before dawn. "Then my father would pinch my ear," a man remembers decades later. The father would say, "Wake up and listen. Let my words enter your ears and your head" (Underhill, R.M., 1939). The mother would do the same thing to the girl. The voice was always a low monotone, the youth always floating on the edge of consciousness. One heard a drone, a hypnotic drone in the desert night. The voice would insist, "Wake up! Wake up!" yet continue in a tone that induced drowsiness. Boys must run and run and run. Girls must grind corn. Do not be idle, the voice would instruct. These messages did not come once or twice or three times; they were always part of the O-otam night, and in time, part of the listener's mind. The admonitions delivered so softly became something very hard: unquestioned rules. "You thought you were still asleep," reflected an adult, "you thought you were hearing that voice in your dreams" (ibid.).

Running was important. A girl must be able to run with a basket or to run from the enemy. A boy must be able to run with a message or to run toward the enemy. The voice would say:

> Listen well to this talk. To all of it listen! Look well at my mouth. Keep looking and you will learn. Thus I say to you, that this night, while darkness covers us, you are not to sleep early in the evening. You are to listen to me.
>
> Perhaps some time in the night, something will arrive. Then if you are a good runner, you will jump up, you will run [after it]. Early in the morning you will get up. Far over there [down to the west] you will sit and listen for something. You will run over there and something which is good-luck-bringing will meet you: a coyote, an eagle, a hawk. Then you will be a fast runner. You will be a good hunter.
>
> You should run constantly, no matter how much hardship it means; even if you are thirsty, hungry, tired. Then truly this will happen to you. You will throw yourself down repeatedly [worn out]. Then there will come a hawk, an eagle, a moun-

tain lion, a jaguar, a bear and will tell you how to follow
something alive [game] and to kill it.

Then when you come back after much running they will
put before you the clay dish. Then you will merely sit and
look at it. Even perchance it will speak thus to you, "Go to!
Eat me!" Do not eat until a scum has formed on it. If you eat
but a little, you will be a lean man. You will be light. You
will be a fast runner. If you eat much you will be fat. You
will be much wrinkled; you will not be a runner. (Ibid.)

Endurance was the hammer that shaped all sound ideas. The
traditional goal was, "You will be hunger-enduring, thirst-
enduring, cold-enduring" (ibid.). O-otam songs are silent on
aggressive action. Survival, not triumph, was sought. Boys and girls
were encouraged to experience suffering—go without drink, skip
a meal—as preparation for inevitable periods of stress. The young
in Papaguería were not told to take the bull by the horns, or to
dare to struggle, or to strike out alone. The individual depended
upon the group, and the group depended upon caution. There was
no victory over the land, just a desperate truce.

When the boy became a man, and the man seasoned enough for
the nightly council meetings, the voice would continue in a different form. Meetings began with set speeches: work hard; be ready
for the enemy; keep your women grinding corn; run, always run;
show endurance. A fragment from such a talk survives:

Well then! Will you not be ready? Will you not take care?
Already I have said thus to you: that you shall make arrows,
that you shall make bows, that you shall be watchful. When
the enemy arrive, you do not know. It may be at night that he
will come—at night, or in the morning, or when the sun
stands almost anywhere. Beside you do you place your bow
that you may snatch it up and fight? Early in the morning do
you eat, that you may be able to fight? Always I say this. Every
evening I say it to you at the meeting, that you shall keep
near you your bow, your hunting arrows, your war arrows,
your quiver. (Underhill, R.M., 1939)

Repetition produced the adult: a mature woman to work for a
household, a "ripe man." Females toiled ceaselessly at basket
making, farming, food gathering, a hundred-odd tasks. The young
were supervised by the old. Until marriage, usually arranged, the

daughter obeyed the mother; with a husband, she became the vassal of her mother-in-law. Only by surviving, and finally becoming old herself, did the woman gain power. During her prime, she, like the man, was esteemed for muscle. Pot making, mat plaiting, skin dressing, weaving, and betting were the preserve of the aged who were too old for hard work (ibid.).

Fetching water took up much of a woman's time. The average O-otam female spent more hours toting an olla than gathering food. The walk might be three, four, up to ten miles. Some of the drip springs required half an hour to fill a three-gallon olla; a woman might stand in line all night.

Some women rebelled and refused the life. They were called "light women." Always dressed as if ready for a wine ceremony, these ladies wore cotton skirts and smeared their bodies with colors. Red and white clay adorned the arms and legs, black lines underscored the eyes. The men called the look "dizziness." Such a woman would meet a man at a festival and go live with him for a while. Between bouts, she might return to her father's house. She refused to be industrious. "The light woman can't help it," Papagos explained, "her heart bubbles over." Another observed, "When you talk to her, she doesn't know what you say. Her heart is outside her, running ahead, to the next dance" (ibid.). With time, such a woman might give in and settle down to the grind of corn. "Will you take me to your home?" one light woman asked a man. "Will you pick cactus fruit for me, grind my corn, and fetch my water?" he demanded. "Yes, I will pick cactus, grind corn, and fetch water" (ibid.). The light went out.

Men experienced the same supervision by the old. A boy became a man and married, yet still lived near his father and listened to his words. "My son has moved away to get more land but I'll be watching him," one father explained. "As soon as he has any trouble, I'll be right there. I have to help him; he's my child" (Underhill, R.M., 1939). For the male, life was a disciplined ordeal. Boys always ran kicking mesquite balls to keep in shape. Various challenges, such as journeys to get salt and warfare, must be met if one wished to be a ripe man. By cruel necessity, males were forced to harvest an *ohp*, an enemy, generally Apache. The O-otam seemed to hate war but became quite good at it. Boys were raised up on war stories; around age twelve they were taught the use of the bow and the shield. Marksmanship was sharpened tossing bundles of rags into the air and impaling them with arrows.

Defense was made clear by having arrows shot at one. Songs cast killing in an ambivalent light: "It is that vile wretch not we who is uttering threats." Men colored their faces black for combat, a look reserved for three activities: dizziness, drunkenness, and homicide. The ancient chant ran:

> My desire was the black madness of war.
> I ground it to powder and therewith I painted my face.
> My desire was the black dizziness of war.
> I tore it to shreds and therewith tied my hair in a knot.
> (Underhill, R.M., 1939)

Scalps were treasured as cures for sexual and nervous disorders. It was said a scrap of hair from another human's head "worked for him like a relative." Such helpers were fed tobacco and eagle down (ibid.).

Given the toil in the fields, the necessity of killing, the treks to get salt, the emphasis on tormenting the body in the quest for endurance, some men fled the way. They became *berdache*: transvestite homosexuals. Suspicious parents usually discovered this leaning by putting the boy in a small hut with a bow and some basket-making material. Then, they would burn the enclosure to the ground. A child that fled clutching basket-making material was a *berdache*. He would grow up to dress and live as a woman. The women found his muscle useful for their work; the men visited him now and again. The obscene nicknames given by *berdache* were treasured. Like the light woman, the *berdache* testified to the hardship of the life by fleeing it (ibid.).

The constant drilling of individuals provided the troops for constructing a community. Kin ties dominated the village and laced one village to another. Several groups would coalesce into one large unit at the fields during the summer rains. Fall scattered them ten or twenty miles away at the permanent water of the well villages. A half a dozen larger groupings had a faint existence on the basis of dialects. But actually Papaguería lacked organization above the village level. War, trade, and hunger sometimes welded the communities together. The dilute resources of the desert constantly drove them apart.

Just as the obsession with discipline sprang from the facts of desert life, so too did the constant motion of the populace and their inability to make a tribe a nation. The challenge of finding concentrated energy is made vivid by the deer. Villages had professional hunters, no more than one or two per community. These men were trained to do the work from the age of ten or so. They serviced the rest of the families, and other men cultivated their crops for them. Steeped in the lore of the deer, devoting their time to the hunt, they averaged twelve to fifteen deer per year. A good rabbit hunter might bag six in a day (Castetter, E.F., and W.H. Bell, 1942).

Slim pickings mandated that individuals act as groups. Papagos, lacking the resources to enforce a garrison state, achieved the same conformity by constantly meeting with each other. Summer gatherings were held in a round building, the Big House; with winter camps at the springs, men met outside. Only males attended, and only ripe men, humans who had proven their worth in war or some other esteemed activity, dared talk. A Keeper of the Smoke led the discussions; he announced them, made the fire, spoke first. He knew the traditional speeches, and went by many names: the Wise Speaker, the One Above, the One Ahead, the One Made Big, the Fire Maker. The Keeper had a messenger, the Leg. Such leaders were chosen by the group and kept their posts until death. They were not young. In the twentieth century one Keeper was ninety, his protégé and heir apparent, seventy. Meetings were held every night and covered almost anything. "And if there was nothing to discuss," one Papago observed, "they would come together and talk" (Underhill, R.M., 1939).

The men entered the building (about twenty-five feet in diameter) and squatted on their heels, arms folded, heads bent. Hair was tied back to avoid distractions. A man who dozed off awoke with a lit cigarette between his toes. Sessions began with a ritual smoking of tobacco and the set speeches urging moral behavior. Then a loose agenda unfolded. Visitors were greeted, and questioned. Farming, hunting, war, ceremonies, games, and what not were surveyed. A man might report the track of a deer, or a human. The old talked and judged—"It pleases me," or "It does not please me"—the young kept silent. Decisions were voted four times; a man could change his mind. Things took time; Spaniard, Mexican, and American were baffled and irritated in turn by the length of O-otam deliberations. A seventeenth century Spanish account confessed:

> . . . it seems the Lord must have given him words to enable him to speak so long, for he harangued them for a space of

two hours, a thing difficult even for a great preacher. After-
ward another took up the thread . . . In this manner day
dawned upon them and the following night it was the same
. . . (Ibid.)

Forty years ago there was a similar observation. A meeting had
dragged on past midnight; one participant said it was very simple.
"We said," he explained, "that at the festival a horse had disappeared, but it had been found at the house of a man in the next
village who didn't know the owner. Then someone said that some
spurs had been found too, but they had been claimed." His American listener demanded, was that all? "Yes," he patiently continued, "but everyone said what he thought about it" (ibid.).

Decisions had to be unanimous. The burden of consensus made
listeners of the young, who had to learn the tactics of utterance.
The ripe man also groaned under this onus; no one willingly
offered thoughts doomed to limited support. Dissent was masked.
Sometimes the strain for agreement became too great, and a village
would split into several. But this was unusual; the desert did not
offer many spots for new villages (Bahr, D.M., 1964). The
O-otam way produced leaders who could not command and fol-
lowers who could not disobey. Meetings boasted much talk and no
debate.

The same lax but iron hand meted out punishment. Both the
child and the adult experienced little physical correction. Words
answered wrongs. "When people have done us harm," according
to Papago thinking, "they should never know from the way we act
what we feel about it. We should never speak of what they did.
Only we would keep away from them. After a while they would
know" (Underhill, R.M., 1939). Gossip whittled at a person's
standing in the group. Marriage, hunting, farming, everything
grew more difficult if there was talk and people kept away. To be
left alone in the Sonoran desert was to be left for dead. The isolation dreaded in the council meeting was just as frightening to the
O-otam in his daily life. Wayward individuals were lectured by
their families and shunned by others.

Once a man gambled away all his property, and then, in a
passion, gambled away his wife's as well. The second act was
serious. The woman did not attack him openly. Instead, she served
his evening meal by digging a hole and tossing the food into it.
"I suppose," she said, "this is the serving dish you like since you

have taken away my basket." The gambler walked off and climbed a high rock near the village. He cried out and people came from their huts to stare at him. Then he jumped. Such was O-otam jurisprudence.

Imagine the land as a flat sheet rippled by wave after wave of mountains. The sheet holds furrows of dry valleys and almost no water. Imposed on this topography are kinship groups; they act as webs distributing resources among humans. Another web connects village to village. This fragile design etched on the land made consensus a necessity, and criticism intolerable. The importance of other humans was made clear every time a person ate. One family might cook up a pot of gruel every night and dispatch two large spoonfuls to all the people in the village. Another family might make gallons of some concoction and give it all away. A deer hunter would return to the village with a buck. People would come to him with a few beans, or some corn. He would give them a chunk of venison. There was no bartering. It was just done. The recipient of a pot of beans would note the quantity and at some future time return a gift of like amount. To refuse to share, to give back less than one received, "was to commit social suicide" (ibid.).

This sharing of food·was not a ritual gesture, or a now and then occurrence. For Christians, Christmas stands for something; for the O-otam the sharing stood for nothing. It was the way resources were distributed among the people every day of the year. "If a man took my horse without permission and lamed it," one Papago argued, "I would not say anything. I would even lend him the horse again. If I should get angry with him, he would have a right to be angry with me" (ibid.). This response, so baffling to outsiders, reflects the fact that the outsiders have never really been inside the Sonoran desert. Among the cacti and spiny trees, in a place where the rains bring flood or do not come at all, here, the only lifeline worth having is another man. The one thing that can level the oscillations of the desert year are people one can depend on.

The attitude of mutual dependence helped bind village to village. Harvest called for extra hands, and the pay was produce. Kin would be invited and the same relatives might come year after year. The helpers took what they wanted; the chief restraint was the obligation to repay in kind someday themselves. "But those who are stingy can't get anyone to work for them," the Papago noted (Underhill, R.M., 1939). Often the desert O-otam toiled for the riverine Pima; an arrangement between specific families might continue for decades. In this way the constant waters of the rivers were drawn into the dry heart of Papaguería.

Trade was another line tossed between population clusters on the Sonoran desert. When Father Kino was lingering at Kuitatk in 1698 an old woman walked in from the west. Her basket held snails and shells she had bartered near Yuma (Bolton, H.E., ed., 1936). Salt, dried meat, and cactus fruit were traded with the Pimas. But such commerce stayed at a level of crude barter and was largely overwhelmed by the simple giving of gifts (Bryan, K., 1925; Castetter, E.F., and W.H. Bell, 1942). Conventional trade faced two barriers. Unique products were rare and trade was restricted to salt, shells, and the like. But more importantly, the desert made hard bargaining unwise. Since people needed other people the gift was more potent than the trade.

The sharing characteristic of the village was periodically practiced on a larger scale. By holding intervillage games one group's good harvest could be invested in other groups less fortunate. The visiting community usually arrived a day or so early: a messenger, with a bundle of sticks, was sent into the host village with a formal challenge. The Keeper of the Smoke took the bundle of sticks, and each household grabbed one or two. The sticks were the humans camped outside the village; by taking a man's stick a bond was sealed between families. The hosts gave food, the guests entertainment and good luck. "If they didn't give much one year," the Papago explained, "we knew they had a poor harvest and we gave even more when it was our turn. But if they kept on that way, we knew they were stingy and we asked in meeting for another name" (Underhill, R.M., 1939). Sometimes a family would toss out gifts in various directions just to test the winds of exchange. Some records of these transactions have survived. One year a man in Santa Rosa gave the following to a family from Kuitatk: three large winnowing baskets of cooked squash, cooked beans, and ash bread. The next year as a guest, he received winnowing baskets of dried beans, dried corn, six large pumpkins (ibid.).

The wife of the giver handed the goods to the wife of the recipient. The giver's wife would often playfully chase her lobbing tortillas and flinging cactus syrup. The visitors sang and danced. The "Name Song" was a common choice. Running seventy stanzas, it provided abundant opportunity to mention the name of the host. As the verses unrolled and a man's name was called, his wife or daughter ran with some light object in hand. Her counterpart among the visitors chased her. Sometimes a number of women would join the pursuit. Once caught, the woman led her guests to a place where "the value of her husband's name" was found in the guise of corn, beans, or other edibles (Russell, F., 1908). Thus were profits plowed back into other human beings.

Gambling on footraces offered another way of distributing goods. Contests were waged between villages. "We bet everything," a Papago recalled. "Women would take off their dresses, men their shirts. Sometimes when people went home again, they had not a single pot or basket in the house. They would dig holes in the ground to serve food" (Underhill, R.M., 1939).

Village runners trained hard. Daily they ran twenty-five miles; their diet held no meat, no honey, no cactus syrup. The morning workout was preceded by a meal of water. Old men supervised the runners and sang appropriate songs. Help was sought from the spirits. One runner met a hawk: feathers grew from his elbows and wrists. Another athlete became unreachable by carrying a piece of a rainbow. The interest in running seems to have paid off. In the eighteenth century Father Pfefferkorn found that the people of Pimería Alta could outrun a horse (Pfefferkorn, I., 1949).

Wild betting could temporarily gut a village of property. But the chance for greed was illusory. The winner could do but one thing with his bounty: give it away. Generosity won esteem and tied other humans to the giver (Bahr, D.M., 1964). As in other activities of the O-otam society the individual gave way to the group lest he succumb to the desert.

This sharing came from the land itself. The resources were hit and miss, bountiful or absent. Fruit on the tip of the saguaro grew here better than there and some places not at all. Acorns, seeds, roots, and leaves were not evenly distributed. The land that concentrated water also concentrated food. The land that had erratic rainfall had erratic productivity. One village watched the fields die in drought while a neighboring community bathed in life-giving moisture. Risk was the only constant. Share. The man who hoarded, who saved, who said he and his blood would make it on their own through thrift and wise judgment, such a man led his kin to extinction. The Sonoran desert was an enormous casino where a man could get an edge, but never savor control. Power came from toil and could only be stored in other human beings.

The drone of work in the ears of the young, the salt journeys, the killings, gift sharing, races, bets, endless meetings, these were the techniques for acquiring resources in Pimería Alta. Practices varied in detail and emphasis from place to place but they resulted in one single condition across the Sonoran desert: societies of abundance. This is not a clever twist of phrase. Scarcity and abundance are not the simple arithmetic consequences of resources. The way can make the difference. Go to the modern American city and look at people short of food, children short of clothing, districts short of work. This society harboring hunger, privation, and idleness consumes 30 to 40 percent of the planet's resources in tending to the needs of 6 percent of the planet's humans. The shambles called ghettos, poverty pockets, blighted neighborhoods were alien to Pimería Alta. Having little, they shared all. The society of abundance was man's response to the desert.

The land determined the food, the food fashioned the way, the way became the mind. The final price for the O-otam was paid in thought. From the Gila to the Gulf the desert people who learned to share, to toil, to live for the group, hungered for the dream.

Dreams

The man hunted alone. It was long ago, probably before Cortez burned his ships and walked his menace into the heart of Mexico. A deer came to this man, and then suddenly changed into a human. The man followed this creature to the ocean. No one saw him for a month. Then, he reappeared in his village, shy, just like a deer. He entered an empty house and sang all night. His kinsmen milled about; looking in the hut they saw a buck. The deer bolted, the people seized the beast. Gradually, shifting from deer food to man food, the animal melted into an O-otam. He taught the people the songs of the deer which became the way to treat deer sickness (Densmore, F., 1929).

Moderns call such knowledge tales, myths, or rituals. For the riverine Pima, the desert Papago, and the Sand people west of the Growlers, it was simply information. Dreams came only to men, and they were not regarded as blights on a good night's sleep. Aborigines realized that thoughts met on the edge of the workday world might be a clue for secular behavior. The O-otam and the Americans lived in two separate deserts. Societies that possess access to groundwater, petroleum, electricity, and subsidies from other regions can either regard the irregular weather, the striking topography, the unpredictable productivity as aesthetic curiosities, or not regard such facts at all. Confronted with such a maze, moderns can literally bulldoze through it. The Indian could not. Ideas about the desert were valued. Humans still listened to the deer.

A different power base dictated a different response to the Sonoran desert. Rain was not measured in gauges, but beseeched by dance and song. In the wine ceremony at the end of June and in early July, the results of many O-otam dreams were offered to the skies. The event enabled the people to coax the monsoons, and, by tradition, offered a respite from traditional constraints of discipline and civility. Words were spoken in anger, marriage bonds dissolved, and the frustrations caused by group dominance over the individual purged. The saguaro fruit was mashed and fermented into strong drink for a saturnalia.

> Ready, friend!
> Are we not here drinking
> The shaman's drink,
> The magician's drink!
> We mix it with our drunken tears and drink.
>   (Underhill, R.M., 1938)

Launched with the mockingbird speech (because the bird was so eloquent he could "stretch his words like ropes between the mountain tops"), the quest for rain took several days. Women painted their breasts with birds and butterflies. Young men rubbed their soles red so that when they collapsed in drunkenness their feet would gleam in the firelight.

> Dizziness is following me!
> Close it is following me.
> Ah, but I like it.
> Yonder far, far
> On the flat land it is taking me.
> . . . .
> Dizzy women
> Are seizing my heart.
> Westward they are leading me.
> I like it.

One on each side,
They are leading me.
(Ibid.)

The sum of O-otam dreams made for days of song cycles. The
words yearned for rain.

The wind smooths well the ground.
Yonder the wind runs
Upon our fields.
The corn leaves tremble.
(Ibid.)

The highest end of Papago activity was the dream. The salt
journeys led to dreams, war led to dreams, anything, with luck, led
to dreams. Humans that lacked power saw the information in
dreams as an edge. Shamans, those who could diagnose sickness,
devoted their lives to this kind of knowledge. Because their power
came from without the community, such men were dreaded and
sometimes killed. In the mind of the shaman there were two kinds
of sickness. Wandering sickness came from fever, pus, sores, and
the like. Anyone could be struck down by these maladies, even
Spaniards, Mexicans, and Americans. But far more serious was
staying sickness. Only the Indians of Pimería Alta experienced this
condition, *ká:cim múmkidag*. "What is the meaning of *ká:cim
múmkidag?*" a twentieth century shaman asked rhetorically:

Look, something is called like that
of the diversity of our sicknesses,
and it never wanders,
and it is just right here,
whenever it is close to the people.
Look, I will explain something clearly,
which that the ocean lies over there,
well, look, as for that something is quite apparent,
that it is there, and there it stays,
and it never wanders.
It just stays there through the years
for as long as it lies, that one.
(Bahr, D.M., et al., 1974)

The staying sickness had many causes. A butterfly, a deer, a
coyote, a rattlesnake, a caterpillar, a frog could strike a man down.

Other creatures than man had dignity, and when offended, they could retaliate. To this day Papago are still being sickened by such mishaps as running over a Gila monster with an automobile (Fontana, B.L., 1975). The desert was strewn with powerful pathways. Shamans learned them, and negotiated.

For moderns, staying sickness comes from other humans and is neurosis or psychosis or half a hundred titles. It is caused by what other humans say or do. Various modern schools argue over how to drive it away. But the current residents of the Sonoran desert do not respect the deer nor buckle to the ground from the blow of a butterfly. The demonic potency of the natural world has been walled off by resources of industrial life. Groundwater has retired the rain dance. The O-otam and other groups lacked such armor against the desert. And so they sought protection by exploiting other resources. Life became a trek toward dreams.

This journey led to many places. By killing a man and cutting a patch of hair from his crushed head, a Papago warrior ventured toward insight. Returning from combat, the killer is met by an older man who ties his hair and sings of a great yellow buzzard.

> But now you were acknowledged [by killing]:
> You made your request to him and wept.
> Then he picked you up, and far he threw you.
> You fell, half dead; you awoke, you came to yourself.
> (Underhill, R.M., 1938)

The warrior is given a scratching stick because he cannot touch himself. He is isolated from the village because he has too much power.

> There is a tree standing alone,
> Casting a cool shadow.
> Under it you will seat yourself.
> Your child must not come near and look upon you.
> Your wife must not come near and look upon you.
> (Ibid.)

The man sits alone for sixteen days (four times four). Each morning his elderly guardian brings him a cup of corn gruel. The man truly seeking the dream lets the meal settle and takes but a sip of the water. Every four days the cup is smashed. By starving the body of sugar, the warrior reels toward visions. After sixteen days, ceremonies return him to the community of the village. If he has

been favored, he will have seen past the land into a song. He will
have found power.

Eagles were a like source. They must be killed. Usually, this path
was taken by the young. A nest is discovered in summer; once
again, an older man supervises. The youth and the dead bird sit
under a tree for four days. In the swirl of days and nights, the
raptor comes to the boy. His destiny is revealed—gambler, runner,
and so forth. Sometimes a youth kills for several summers in search
of an answer he can live with.

> The Mountain of Reeds
> Stands up at the west.
> There an eagle cries,
> The flat land resounds.
> (Ibid.)

Another way of knowledge is the salt journey. Four days each
way, the trip is hard. Those who take this path may win dreams for
themselves, and rain for their people. For success, the trek must be
made four successive years. The travelers run carefully lest they
hurt some animal and offend its dignity. They sleep with heads
toward the ocean so that "its power can draw them on." Only three
waterholes link the heart of Papaguería with the sea. After the last
one in the Pinacate country, the salt seekers go a day without drink.
The trip home is silent and the body may not be touched. Upon
return to the village, sixteen days of isolation is mandatory. "The
ocean water hurts my heart," according to the ancient salt song. For
a reason:

> Grant powers to me:
> Great speed in running.
> Great industry.
> Great skill in hunting.
> (Ibid.)

These ordeals, and there were others, brought information into
a world largely closed to innovation. The knot of discipline could
be loosened, perhaps cut, by dreams. New ways could be ventured
to live a life, to bring the rain, to beckon the deer to the bow. They
constitute a dialogue with the desert. This has now all but ended.
There is not much out there worth knowing if humans no longer
live out there.

This was the land of the stick. The ephemeral vegetation racing through germination, leaf, flower, and seed during the rainy season were the only bits of life free of subsurface water to any degree. For the rest—the man, the mesquite, the deer—the year must be traversed on the remnants of rain buried in the earth. Perennial trees acted as pumps, succulent cactus as water towers. The pace of life was in the grasp of things without hands—leaves, meteorological highs and lows, the metabolisms of various herbivores, the habits of rats. Humans lived in concentrations decided by the cells of plants and animals. Groundwater rates were the province of drip springs and tree roots. In such a world, humans will listen to coyotes or eagles in hope of getting some leverage to alter the flow of resources. Bodies will be tortured for a scrap of information. Dreams will not be dismissed.

The only escape from this world is energy—energy to make the materials concentrate and cycle more swiftly. The people of Pimería Alta seemed to sense that such a change would some day come. Elder Brother, a sort of Moses in their legends, left a prophecy first learned by whites in the 1920s from the Gila Pimas. The prophecy said a "they" would come. The Piman people were fated to survive so that they could witness the acts of this "they." A time would come when things were pushed, when the pace was quickened.

> And you will not be the ones to kill the staying earth,
> I will leave it to them,
> And they will do it.
> And these will kill the staying earth,
> And even if you don't know anything,
> And you will just be feeling fine,
> And you will see it
> When it happens.
> (Bahr, D.C., et al., 1974)

3. Shovel

His food lacked salt. He sought vile herbs for seasoning. He did not smoke, he did not take snuff, he did not drink. He did not sleep in a bed. Two coarse shirts sufficed for a wardrobe, the sweat-blanket off his horse for the night. But then he slept very, very little. Prayer took up his spare time, and when he prayed he wept. Or he craved penance. "One night," his companion of eight years recalled, "I casually saw someone whipping him mercilessly" (Bolton, H.E., ed., 1919). University educated, he entered the Society of Jesus after surviving a serious illness. His name was Eusebio Francisco Kino. Between 1687 and 1711 he opened up Pimería Alta to European men, ideas, and ways of using resources.

He traveled some 75,000 miles in bringing plows, horses, iron, and Catholicism to the land of the stick. The Sand Papago, the two-village O-otam, the riverine Pima all knew him. He is said to have given the Gila Pima some blue beads; at the turn of the twentieth century a secluded Pima shrine gleamed with a strand of blue beads (Russell, F., 1908). No one remembers clearly what he did. Today there is an annual religious festival at Magdalena, Sonora, where he is buried. The Papago and Pima always come. But the Indians pray to a patron saint not Kino's and make no mention of him. Yet they have trekked to this small town (where he died two centuries ago) for as long as anyone can remember. The link, like Kino's impact, is lost but still haunts any understanding of the Sonoran desert.

There is no certainty about what Kino looked like, what his nationality was, how he spelled his name. Still, he left some records to suggest what he was about. The father wanted to bring Baja California within the blessings of the church. For years he had the Indians at Caborca toil on a boat in the middle of the desert. This ship, built by humans who had probably never seen one on the water, testifies to Kino's greatest attribute: will. He never let the facts of the Sonoran desert stand in his way. The "very fertile and pleasant lands and valleys of these new conquests," he wrote in his memoirs, already have "very rich and abundant fields, plantings and crops of wheat, maize, frijoles, chick-peas, beans, lentils, bastard chick-peas, etc." The place was studded with "good gardens" and vineyards; cultivation of sugar looked promising. Besides "all sorts of garden stuff," Pimería Alta held fig trees, quinces, oranges, pomegranates, peaches, apricots, pears, apples, mulberries, pecans. Timber for construction was "very good." For beauty, one could pick "Castilian roses" (Bolton, H.E., ed., 1919).

Grass seemed without end. Kino pushed stock raising; he sketched a Pimería Alta filled with ranches hosting "cattle, sheep, and goats, many droves of mares, horses, sumpters—mules as well as horses—pack animals necessary for transportation and commerce, with very abundant pastures all the year, to raise fat sheep, producing much tallow, suet, and soap, which already is made in abundance." The climate featured "neither too great heat nor too great cold." Hints were dropped about gold and silver (ibid.).

He found the native Piman people hard working and hungry for God. This was a blessing because the neighboring Apaches seemed to love war and shun the cross—they would not be broken until the end of the nineteenth century. The priest envisioned a budding economy based on exports from this land. The Piman people made "very good fabrics of cotton and of wool; also, nicely made baskets, like hampers, of different sizes, many colored macaw feathers, many deer and buffalo hides, and toward the sea coast much bezoar, and efficacious contrayerva (*Dorstenia contrayerva*), and in many parts the important medicinal fruit called jojoba" (ibid.). Perhaps, salt, fish, shrimp, and oysters could be exploited on the coast. Kino saw the missions reducing the Indians, and the reduced Indians reducing the desert. All this resulting from the power of God, plows, iron, horses, wheat, Castilian roses.[10]

Kino's misunderstanding of the Sonoran desert enabled him to act. Blind to the way its resources were distributed and unimpressed with native efforts to harvest these resources, he readily pushed for plows, wheat, permanent villages, new herbivores. The Piman people were easily comprehended as amiable incompetents. The O-otam, eating beans, became the Papavi or Papago: bean eaters. The riverine groups were soon called Pima—a corruption of an expression in their language meaning, I do not understand (Russell, F., 1908; Underhill, R.M., 1939). "They gave us only beans," one soldier complained of the O-otam, "for although they have an arroyo and rich land with a ditch to irrigate, they are lazy and do not get enough corn for the whole year" (Underhill, R.M., 1939). "Naked and poor," the desert people were found markedly "inferior" to the river tribes. "They have no water," noted the Spaniards, "but live on the desert" (ibid.).

A late eighteenth century visit of the Papago left this description:

> The Papago Indians who dwell in them [brush huts] live in greater misery and poverty than can be imagined. Seldom have they any fixed habitation, because driven by necessity, they wander almost continually, sometimes in the sierras to maintain themselves by hunting deer, wild sheep, rats and certain roots; at other times, in the arroyos to live on the bean of the mesquite, and again in the plains, seeking pitaya, the tuna, the saguaro and other wild fruits. Sometimes they go to the Colorado River to obtain food amongst the Yumas through their skill in their disorderly dances ... (Castetter, E.F., and W.H. Bell, 1942)

Certain of their superiority, the Spaniards were appalled by any resistance to the new ways.[11] Kino soon confronted protest from one group of Indians. They gave four blunt reasons. The people believed the holy oils killed; also, they were repulsed by the European custom of hanging humans. Besides these trifles, they did not like the new ways of using resources. The priests, they argued, "required so much labor and sowing for their churches that no opportunity was left the Indians to sow for themselves." Also, the tribesmen believed that the Spaniards "pastured so many cattle that the watering places were drying up ..." (Bolton, H.E., ed., 1919).

For Kino, such talk was simply the work of Satan. Like the Americans who came after him centuries later, he could not conceive that the desert O-otam, riverine Pima, and Sand Papago could have any rational grounds for clinging to their ancient ways. And yet the Indian complaints down through the years have a grim consistency: the newcomers to the society of abundance are attacked for taking too much from the land for themselves and their animals.

Always one returns to the land with its meager hydrologic budget. The Indian way was motion; the European way was sedentary. Adobe was recommended over brush huts, farming over foraging, the cow over the deer. Water-rich pursuits were stressed. All this had consequences. The population of the Pimería Alta which Kino found dwindled; it did not regain its seventeenth century size until the mid-twentieth century. This charnel-house effect was typical of

Spanish efforts at converting Indians (Hastings, J.R., and R.M. Turner, 1965). In part, humans died from new diseases; in part, from new ways.

Kino brought more than the cross. He distributed wheat, chick-peas, bastard chick-peas, lentils, cowpeas, cabbages, lettuce, onions, leeks, garlic, anise, pepper, mustard, mint, melons, watermelons, cane, grapes, roses, lilies, plums, pomegranates, figs, cattle, oxen, horses, mules, burros, sheep, goats, chickens, iron. Three of these introductions were probably crucial. As the Spaniards advanced from Mexico City north, cattle numbers grew as human numbers collapsed. Herbivores replaced omnivores. By 1694 the headwaters of two rivers in Pimería Alta, the San Pedro and Bavispe, held 100,000 head. In 1703, Kino kept 3,500 head as a reserve for starting new herds in the desert (ibid.; Bolton, H.E., ed., 1919). Cattle probably competed with native herbivores and with humans for range. The O-otam, most of whom did not take up herding until the late nineteenth century, seem to have hunted them like deer.

Wheat was another major introduction. The riverine people lacked a winter crop; the stream flow could not be utilized by them during those months. Wheat (and barley) filled this gap. Quite simply, the ability to collect solar energy through agriculture was probably doubled by this plant. This is an explosive increase. Food is energy; modern societies are testaments to the effects of rapidly increasing energy flows. Finally, the Spaniards brought the horse. The desert people, who controlled no animal but the dog, learned mastery of a beast weighing half a ton. Transportation of goods and humans was instantly expanded. Distances shrank.

Kino's innovations were a triphammer in the Sonoran desert, and as the years revolved from cactus harvest to cactus harvest these crops, animals, and ideas beat remorselessly on the old patterns of water use and land use. They presented the Piman people with new ways to concentrate and mine resources. The cow looted the vegetation, the horse shattered ancient tribal boundaries, the wheat created sudden new pockets of energy. These Jesuit ingredients for a new Christendom of missions, sedentary Indians, and a European economy upset the old farming and gathering world of the desert and brought two hundred years of turmoil to the humans of Pimería Alta. While the Spanish influence ebbed through the eighteenth century, and the missions retreated to a few river outposts, the natives were left to wrestle with the implications of these factors. They entered an era of constant warfare.

Kino's plans for the Sonoran desert drew the Apaches. A raiding society occupying the mountains of central Arizona, the
Apaches stole horses, humans, and food. From the 1690s until the
1880s, they pillaged Pimería Alta. By the eighteenth century, the
Pima villages on the Gila were enduring raids every few days.
Following natural corridors like the Santa Cruz and the San Pedro,
the Apaches would hit the trail for sixty or ninety days and penetrate hundreds of miles to the south. Besides the almost daily rob-
bery, large war parties visited about twice a month. The Spaniards
slowly let go of their new domain; the Pima surrendered the San
Pedro valley and contracted their lands on the Gila; the desert
O-otam remained within their arid bastion (Hackenberg, R.A.,
1955).

"We asked the Indians why they live so far from the river,"
observed a priest visiting the Pimas in 1775; "they replied that . . .
by living apart from the river they were able to have a clear field
for pursuing and killing the Apache when they came against their
town" (ibid.). Father Pfefferkorn recorded this menace with feel-
ing.

> At present . . . Sonora is not a shadow of what it was, and
> there remains for it only the remembrance of its former
> prosperity . . . These savages have for many years raged ter-
> ribly in Sonora, have cruelly murdered or carried off into
> captivity a large number of Spaniards as well as converted
> Indians, have stolen an indescribable number of horses, mules
> and cattle, and have committed other like devastations.
> (Pfefferkorn, I., 1949)

Another priest lamented, "Sonora is on the verge of destruction,
for we have seen . . . scarcely ten places inhabited by Spaniards,
while there are more than eighty ranches and farms destroyed by
the enemies" (Hastings, J.R., and R.M. Turner, 1965).

Lack of records makes the dimensions of this carnage uncertain.
Two things appear evident. Kino's work created new concentrations of humans, of crops, of cattle, and of horses. Almost instantly,
the Apaches took to seizing these new concentrations. This economy of blood made the raiders a legend, and by driving European
observers from the area, made most of the history of Pimería Alta
for over a century a blank page.[12] Bits and pieces are known. Late
in the eighteenth century the Gila Pimas began making wooden
imitations of shovels and plows. Cattle seemed to have diminished

with the Spanish retreat. Apaches stole herds from the south and drove them home. But the riverine farmers on the Gila did not get their own stock until around 1825 (Hackenberg, R.A., 1955). A 1746 reference describes wheat fields among the Pima. In 1774, a Spaniard claims a field on the Gila so large a man in the middle could not see the ends. A year later, a priest states, "In all these [Pima] pueblos, they raise large crops of wheat, some of corn, cotton, calabashes, etc., . . ." (Castetter, E.F., and W.H. Bell, 1942).

Following Kino's death in 1711, the desert Papago all but drop from view. There are hints that the cow slowly replaced the deer, that melons and oddments from European gardens found some place in the tradition of akchin. Apache raids shrank the number of O-otam villages; the desert people began collecting patches of hair from Apache heads. When Lt. A.B. Chapman of the American army visited them in 1858 he noted that the Papago "seem to have selected these locations for their towns on account of the Apaches. To them, now, the Apaches cannot approach without crossing extensive arid plains" (Hackenberg, R.A., 1964). By the early nineteenth century the tumult drove the O-otam to a singular invention: they began notching calendar sticks.

All in all, Kino's efforts seem to have had more immediate impact on the humans of Pimería Alta than on the land itself. Cattle numbers were checked by predators, human and wild. Crops permitted some strengthening of sedentary tendencies among the riverine people, but foraging continued, and the Apaches gathered any surpluses, either human or cereal. Horses offered food and transportation, but once again, expansion of the herds was limited by exports to Apachería. In short, the economy that Kino dreamed of with exports for other regions came to pass, but the trade routes did not lead to Mexico City or Europe but to the wickiups of warriors in central Arizona. Societies based on renewable resources persisted with the real changes coming in the way resources were concentrated and distributed.

Records return to Pimería Alta with American efforts to steal the land from Mexico. Soldiers crossing to California stumbled upon the villages of the Pima. "The Pimas and the Maricopas [refugees from wars with the Colorado River Yumas] are wonderfully honest and friendly to strangers," penned Colonel Cooke in 1846. He discovered that they raised "corn and wheat which they sell cheaply for bleached domestics, summer clothing of all sorts, showy cotton handkerchiefs, and white beads." He noted their cattle and mules, and left them some sheep for breeding (Hackenberg, R.A., 1955). Later that same year Cave J. Couts traveled through eighteen miles of villages and fields along the river. "Might be called an Indian City," he suggested. The farmland struck him as "a series of the finest fields I ever saw . . ." (Fontana, B.L., 1957). The Gila was a permanent stream; in the 1840s three men and a family floated a sixteen-foot boat down it to the Colorado (Ross, C.P., 1923). The gold rush of 1849 sent possibly 60,000 Americans to California by way of the Gila. The Pimas fed them, and by 1852, they were complaining of the strain of this charity (Hackenberg, R.A., 1955).

Surprise and praise marked American encounters with the Pimas. Lt. Col. J.D. Graham in 1852 mentioned fertile land "producing crops of cotton of the first quality." Another traveler in the fifties estimated the fields as twenty miles long and one or two wide. H.H. Hutton believed the croplands could easily feed 6,000 humans (Fontana, B.L., 1957). The desert O-otam, less frequently reported, were observed in Tucson in 1859. They seemed "very prosperous," and each had fifteen to twenty-five dollars, some cattle and horses. They occasionally hired out to Santa Cruz valley farmers and made money hauling salt from the sea—"for which they find a ready market in Tubac and Tucson" (Hackenberg, R.A., 1964).

For the Pima, this image of the peaceful garden was about to change. They had spent more than a century assimilating European innovations. Simple things, like the shovel, had trickled into their society. Gradually, the digging stick had widened. The Pima would select a piece of heartwood from a cottonwood tree. From the fire-felled tree, they would rough-shape the wood with more fire. Finally, the blade was finished with a rough stone. The result was something like a tennis racket with the working end about eight inches wide (Castetter, E.F., and W.H. Bell, 1942). The American intrusion in the Southwest brought more changes at a faster pace.

At first the Pima were eager. When some troops stopped at a village on the Gila in 1858 the chief, Juan José, offered their commander $3 in gold for any shovels or axes they might possess. Informed that regulations forbade such sales, he upped his bid to $5 in gold. Still denied, they branded the Americans a nation of liars (Hackenberg, R.A., 1955). A year later when the United States Congress bestowed $10,000 on the Pima as a reward for their crops and Apache killing, the Indians knew exactly what to do. They ordered and received 444 axes, 706 butcher knives, 516 hoes, 15 plows with harnesses, and 618 shovels (ibid.).

It was very simple. The shovels, axes, plows, and other devices of biting iron meant that the earth could be raked and mauled more extensively, water could be diverted to more land, and crops could be increased. The outside world that brought tools to the Pima also brought a market. In 1858, the Gila Pima sold 100,000 pounds of wheat. Next year they peddled a quarter million pounds. By 1860, sales reached 400,000, then a dip to 300,000, and finally in 1862, 1,000,000 pounds. That same year the villages fed 1,000 federal soldiers for one month (ibid.). The stick gave way to the shovel. Humans reached out with iron hands to channel more and more of the desert's resources into their own appetites. The voices in the night that had molded and propelled Pimería Alta to unremitting toil found something commensurate with their ardor for labor: American commerce.

The American Civil War in Arizona was fueled by energy gathered along the Gila. While small bands of humans wearing blue and gray wandered about seeking combat, the Gila Pima powered this miniature war with wheat, corn, melons. The Union army wracked its brains seeking some commodity that would tempt the Pimas to trade. Lt. Col. Joseph R. West ordered 13,000 yards of cloth for this purpose. If these goods failed to stimulate commerce, he contemplated piracy (Fontana, B.L., 1957).

In 1863 troops devoured a million pounds of wheat and a quarter million pounds of corn, cotton, sugar, melons, and beans. One witness thought the Pimas could increase their yields "almost indefinitely" (ibid.). Pima productivity seemed boundless. A visiting general in 1867 described this Gila:

> Their fields were green with fast-springing wheat and barley. In addition they have corn, beans, melons, etc., and have horses and cattle in considerable numbers. One drove of their livestock, over 2,000 head, passed down the road just ahead of us . . . We were told they had many more. The year before this these Indians had raised a surplus of wheat, corn, amounting to 2,000,000 pounds, besides a large supply of barley, beans, etc. (Hackenberg, R.A., 1955)

With the Civil War ended, the Pimas and the Americans made one final adjustment in the way goods were distributed in the Sonoran desert. A few short weeks after Robert E. Lee offered up his sword, Jonathan S. Mason, Brigadier General of the Arizona Volunteers, ordered 600 red shirts, 600 blouses, 600 pairs of coarse pants, and 600 yards of coarse red flannel (the "last will answer instead of a hat"). General Mason was setting forth with 200 Pima warriors to kill the Apaches (ibid.). By the 1870s they were broken as a force in Arizona. The raiding society made possible by Kino's work disappeared from the desert.[13]

The Pimas had won at every hand. Wheat, horses, cattle, and vegetables had reached them beyond the hold of Spanish discipline and Jesuit organization. The Apaches had been endured and finally destroyed. The American army had provided a market for their new skills in agriculture. Congress in 1859 had created a reservation where their villages lay on the middle Gila. The enormous drainage of the river delivered them just enough silt and water. True, the Pimas had sought a much larger reservation since they sensed how vast a section of earth it took to service their fields. But on the whole they were victors in a process succeeding centuries called conversion, then destiny, then progress, and finally, development. By 1865 the Pimas had learned how to get more from their ground without maiming it, how to seize more from the desert's hydrologic budget without unbalancing it.

The desert Papago, claiming land without a living stream, absorbed less from outsiders and faced fewer challenges. Travelers crossing their range toward the California goldfield often died of thirst. Ranchers who encroached lost cattle to Papago thievery and the desert's irregularity. Mines flourished briefly, then collapsed as the veins of metal gave out. Sometimes no rains or bad crops drove the O-otam to the river valleys for work. But in the main, they were left to their own power system of songs beckoning clouds.

This state was not to last. As early as 1857 a geological report had been filed on the Gila valley. "Only a small portion of the valley is under cultivation," the study concluded, "but it is susceptible of being made productive much further away from the stream . . ." (ibid.). Elder Brother had said that everything would be just fine. When it happened.

4. Tractor

"They have very little water here," an Englishman wrote home in 1876 of southern Arizona. "I have heard a man describe Arizona like this, he said: 'Get a box of sand and in one corner put a thimbleful of water and in the other a horned toad and you have Arizona'" (Fontana, B.L., ed., 1965). A year later he said it another way: ". . . take a man who is unbiased in opinion and he will not speak so well of this country if he speaks the truth" (ibid.). This naysayer, Herbert R. Hislop, helped found one of the greatest ranches of the region, the Empire. His assessment of the area was a minority position. Like the Spaniards before them, most Americans saw what they were looking for rather than what they found. During the closing decades of the nineteenth century the citizens of the United States filling up the western half of the country had great difficulty locating one thing: the arid lands.

"The Great American Desert: Where is it?" asked L.P. Brockett in his 1,300-page guide, *Our Western Empire or The New West beyond the Mississippi* (1882). Addressing himself to the Great Plains, the Southwest, and the Great Basin, he explained that wherever whites settled with bovines and plows, the rains increased. Arizona Territory was no exception. Contrary to some loose talk, Brockett knew the region had plenty of water. Rains averaged 15 to 24 inches per year. The river valleys were made for irrigated agriculture; and ". . . most of Arizona is arable . . ." As a clincher he pointed out that "fruits are comparatively plentiful and cheap." Granted a few remote spots may be treeless and short of rain, but these exceptions "may be rich in fossils or other precious metals." There simply was no Great American Desert (Brockett, L.P., 1882).

The Arizona legislature hired another man to write a similar such book. Patrick Hamilton's volume, *Resources of Arizona*, argued, "The water supply of Arizona is sufficient to irrigate nearly all the arable land within her borders . . ." Admitting that the climate of the Sonoran desert was "inclined to be a trifle hot for about three months," he countered this awkward moment of truth with the announcement that the heat was dry and sunstrokes vir-

tually unknown. What lands could not be farmed were ideal for
ranching. Arizona could not be overgrazed. "Ranges," Hamilton
confided, "over which cattle have roamed for years show no fall
off in quantity or quality of the feed. In fact, it is claimed by some
that the ground is enriched by the cattle, and that the native grasses
attain a stronger growth after being pastured for a few years."

Two common questions about the region were easily dismissed
by the author. The Apache menace demanded one solution: "ex-
termination." As for any groundwater problem, there simply was
none. The desert aquifers "are the great cisterns where Nature has
stored a treasure almost as great as the gold and silver in her rocky
hills, a treasure which has long remained unused and unknown,
but one which the cunning of man will soon utilize to make many
a barren waste smile with verdure." Also, most of these aquifers
were under artesian pressure and "inexhaustible" (Hamilton, P.,
1884).

These statements come under the heading of boosterism, lies
justified by greed. They were challenged in the late nineteenth
century by men such as John Wesley Powell, head of the U.S. Geological Survey.[14] They were repudiated by the land itself. Neither
reply broke the power of the original vision. The humans who
confronted the Pima and the Papago in the late nineteenth and
early twentieth century believed the desert either nonexistent, or at
worst temporary. These beliefs had consequence. Since the land was
not really arid, water could be used with abandon. Cattle could be
grazed without thought to numbers. Rivers could be exploited
without concern for other users. For the Americans the problem in
the arid West, and in the part called the Sonoran desert, was not
any shortage of resources, but rather a failure by earlier inhabitants
in the area of technique. What was needed was know-how.

By the end of the Civil War 9,000 whites lived in the Arizona
Territory. "The laying of the iron rail marks the brightest epoch in
Arizona's history," was Hamilton's comment on the penetration of
the Southern Pacific in 1878. "Before the advance of the locomo-
tive," he explained, "the barriers of isolation were removed and
the last vestiges of savagery swept aside. Population increased
rapidly . . ." (1884). Arizona held 40,000 non-Indians by 1880.
Cattle migrated to the area at an even swifter rate.

Only 5,000 head browsed the Territory in 1870. One outfit,
Hooper and Hooker, brought in 15,000 head in 1872. By 1880,
one river valley, the San Pedro, hosted 8,000 cows and 12,000

sheep. The year 1880 found about 20,000 beeves south of the Gila. A few years later the governor reported that "every running stream and permanent spring were settled upon, ranch houses built, and adjacent ranges stocked" (Hastings, J.R., and R.M. Turner, 1965). Non-Indians reached 60,000 by the nineties. South of the Gila, the official cow census was 315,000. Probably, the actual figure was double that; in 1891 the governor believed 1,500,000 cattle grazed in the Territory. Then something happened. During the last decade of the nineteenth century with the range cropped down, arroyo cutting set in across the Sonoran desert. Rivers like the Santa Cruz and San Pedro became dry beds. Swamps and marshes disappeared. Beaver deserted the valleys, and malaria left with the water. The indestructible grasslands that charmed Father Kino and excited Patrick Hamilton were destroyed.[15] The Great American Desert had said hello.

Gila River

The Pimas looked for messages in dreams. This time the announcement came from the river. The Gila flooded in 1868. Then, for five years, it ran low and proved inadequate for watering the Indian fields. A chief, Antonio Azul, was amazed because he thought in the past the river had never failed more than one year in ten. The new iron plows, hoes, and shovels were as useless as the land itself when the water stayed away. Songs did not work this time. The ditches of non-Indians were diverting the river above the Pima lands (Hackenberg, R.A., 1955).

The problem at first was at least tangible. Cattle looting the Upper Gila of vegetation had augmented the flood. American and Mexican farmers seizing the river's flow had kept water from getting downstream. The Pimas reacted almost instantly to this movement of resources into alien hands. They harvested a poor crop in 1869. The Pima scouts and soldiers returned from killing Apaches with the Arizona Volunteers. The first Pima assault against settlers came in November, 1869. Four hundred tribesmen walked onto the corn and bean fields of Mexicans just above the reservation. The Indians picked the crops and kept them. They explained to the Mexicans that both the land and its produce belonged to them. They demanded rent (ibid).

"On the 25th of April," reported the Tucson newspaper in 1873, "the Pima Indians were guilty of an outrage at the house of Patrick Holland, one and a half miles above Florence . . . As everyone here on the river knows, the Pimas are continually trespassing upon the farms of our people and are generally doing pretty much as they please." Traditionally, the Pimas had foraged in the area where the whites now farmed. When they found white farms in their path, they stomped through the growing crops. This angered the farmers. Holland had caught a man and a woman cutting through his barley field. He ordered them to turn back; he showed them a way around his field to the river bottom. The Indians said they would go where they pleased. Holland beat them with a stick.

A while later Holland was visited at his home by sixty Pimas. They had guns and they had bows. They pointed the guns and bows at him, and said, "Now, drive us off!" But as the newspaper cautioned, "Of course, Mr. Holland could only submit, for if he had resisted them in any manner he would have been murdered." Such incidents, the article continued, were not rare. The Indian agent did nothing. Only four years before the Pimas had nearly beaten a Mexican farmer to death. This man had ordered the tribesmen "off his land when he caught them stealing corn . . ." (Fontana, B.L., 1957).

The Pimas did not read the books written by white people. They did not know that Arizona had plenty of water; they did not know the grasslands could not be overgrazed. During the last quarter of the nineteenth century, they dispersed, seeking new concentrations of water. Gila Crossing was founded in 1873–74, Maricopa in 1887–88, Santan in the seventies, and so forth (Hackenberg, R.A., 1955). But American appetites for resources made hopeless any Pima tactics of adjustment. Between 1880 and 1900 three thousand Mormons occupied the upper Gila, founding Safford. Just above the Pima lands the town of Florence boomed on a diet of wheat, water, and silver.

"The Gila flouring mills of Florence," the Tucson newspaper reported in 1875, "presided over by the genial and accommodating Hon. P.R. Brady, are in full blast day and night. The abundant supply of water keeps the machinery in clock working order, turning out thousands of pounds daily of the veritable essence of life . . ." (Fontana, B.L., 1957). One white farmer, R.C. Brown, left a memoir of this era in Florence. "Thought I had plenty of water . . . ," he recalled. He planted alfalfa, fruit trees, and watermelons. A neighbor diverted his water at night. Birds ate all the cherries. Badgers feasted on the watermelons. "Bought some lots in town," he remembered with bitterness. "Florence lots were worth more at this time than anywhere else in the Territory but when I left I could not dispose of them for love or money." Zeal for Florence residence diminished for a simple reason. The town waned because Mormons at Thatcher and Safford stole the river from Florence. By 1904, the Pimas had only 7,000 acres of cropland, and this land was productive only at those times of year when the whites above the tribesmen could find no use for the water (Castetter, E.F., and W.H. Bell, 1942). As Patrick Hamilton had explained in his 1884 text on Arizona's resources, the Indian "farming is of the most primitive style . . ."

The Pimas, described as Arizona's only laboring class in 1865, were perceived as murderers and thieves, and then as lazy Indians, as the water left their lands. The river died before their eyes. The verdant banks of the Gila, which had bewitched and succored tens of thousands of Americans during the gold rush of 1849 and the Civil War, went dead. The river left its clean channel and began to wander. It grew wider and wider until it did not look like a river at all, but rather a sandy waste. The legendary Pima farmers who fed the hungry and traded wheat for iron followed it into waste. By the turn of the century, when Harvard anthropologist Frank Russell studied them, they had reached that state typical of North American natives: enduring ruin.

Thereafter, attempts were made to put them back together. The Presbyterian church won many converts; a tribal revival was held in 1906. The Pima ignored the theology of predestination and demanded sermons on the value of hard work and good behavior. This first decade of the twentieth century was called the black decade by the tribe. Whites attempted to move the Pimas to a new modern irrigation district based on groundwater; the aquifer was alkaline and commercially worthless. For this treasure the Pimas were to give up their reservation lands. Protests finally stopped this swap (Hackenberg, R.A., 1955).

Some placed hope in the education of the young. Pima youths were shipped off to boarding schools in Phoenix. "I left the reservation in 1911 and came back in 1929," one such student recalled. "I finished Indian school in 1922 . . . I worked as a chauffeur and in a hardware store, and as a caretaker. For three years I worked as an undertaker. Quite a few people like me did those things" (ibid.).

Some of the educated Pimas tried modern agriculture on the reservation. They founded the Cooperative Colony and the Progressive Colony. This was unusual since most Pimas then (and many now) believed the native system of farming superior. The colonies were based on seepage water that was too alkaline for successful cropping. A dam built upstream by whites in 1928 removed even this feeble base. Perhaps, some thought, private property would do the trick. Pima land was broken into individual allotments. Things did not go as expected. One man died in 1926 leaving 20 acres. By 1936, seventy-three heirs claimed the land. Eight of these people held shares of 87/573,300: eleven square feet had it actually been divided (ibid.).

The Coolidge Dam completed by the Bureau of Indian Affairs in 1928 was the ultimate panacea. By trapping the Gila above the reservation, it would guarantee water for the fields. The dam never captured enough flow to water the fields. But it did stop just enough water from going downstream to radically lower the water table on the reservation. The villages and colonies based on seepage water after the theft of the river in the 1870s went under (ibid.). As the resources of Pima life were strangled over the decades, the life continued after a fashion.

Until the First World War, the ancient world of praying, planting, and harvesting endured. The touch, feel, and pace of this life still seemed remotely valid. The war sent Americans to Europe and the Americans drove machines and the machines rode on rubber tires. The tires had sinews of cotton cord. In 1917, the Goodyear Tire Company leased land just north of the reservation to grow long staple cotton for this war cord. The new crop spread into the Salt, Gila, and Casa Grande valleys. The whites had hit a bonanza. Pimas picked up some work as laborers. But the cotton farms proved too much for Pima ideas about the world. Prior to Goodyear's investment, white and Indian farmers on the Gila had eked out a living with an almost subsistence agriculture. Both followed horse-drawn plows, both looked much alike working the land. The Indians could keep up in a way more crucial than simple measurements of bushels and pounds per acre would indicate. They could believe in the utility of their efforts.

"Then," in the words of the government farmer assigned to the reservation, "came the tractor. They saw white men sitting up there on top of the tractor and doing in one day what it took them a

week to do with horses. It took the heart out of them." Their
holdings were too small to finance the purchase of machinery;
"they just sat around and waited for the Indian Service to come
and farm it with a tractor for them" (ibid.). They made the move
from a life based on living in balance with their land's resources to
the life of white society. One Pima explained decades later:

> Pimas don't live like Robinson Crusoe anymore. They need
> cash crops and can't live off subsistence farming. Those people
> who gave us the San Carlos Project [Coolidge Dam] meant
> well, but they thought the Pimas were the same in 1930 as
> they were in 1850; all they needed were a few squash, beans
> and corn . . . Today the average Pima needs 60 acres and all
> the water that he can use to make $2,000 to $3,000 from agri-
> culture. This would only be with a series of grain crops, not
> just one crop of barley. (Ibid.)

Today the reservation on the Gila looks like desert. A four-lane
interstate cuts across it. South of the dry river is a new cultural
center where Pimas sell baskets and pottery. Just north of this
building lie the ruins of Snaketown, a Hohokam farming town
abandoned centuries before Kino. Recently declared a national
monument, the dead community promises to be an important job
source for the tribe. Limited agriculture persists. Cash labor off the
reservation brings in money. Federal programs of health, food, and
education draw resources to the Pima. Their numbers blossom,
their physical presence is not threatened. Their use of the land is
gone. They live on a patch of ground by a dead river, but they do
not live off that ground.

The only laboring population in Arizona in 1865 buys its food
a century later. The men, the women, the children feed off storages
of nonrenewable resources. The gutted Gila is a testament to this
new way. One Pima remembers it like this:

> Now the river is an empty bed full of sand. Now you can
> stand in that same place and see the wind tearing pieces of
> bark off cottonwood trees along the dry ditches. The dead
> trees stand there like white bones. The red-wing blackbirds
> have gone somewhere else. Mesquite and brush and tumble-
> weeds have begun to turn those Pima fields back into desert.
> Now you can look across the valley and see the green alfalfa

and cotton spreading for miles on the farms of white people
who irrigate their land with hundreds of pumps running
night and day. (Webb, G., 1959)

Papaguería

Nothing, it seemed, could touch the arid core of Pimería Alta. The
Spaniard, Mexican, and American each came and each left. There
was plenty of time to fit new animals, seeds, and ideas into the
ancient way. The cow replaced the deer, new crops crept into
akchin, Catholicism was molded by native hands into something
finally called Sonoran. While the Pima experienced first prosperity
and then ruin, while the Sand Papago became a new game animal
and were exterminated, the desert O-otam followed their own
path. The twentieth century dawned and still they thrived without
wells, electricity, petroleum.

White would penetrate Papaguería and then retreat. Ranchers
lost their cattle to the O-otam and the desert. Mining camps briefly
boomed (one had a temporary population of 11,000), then collapsed. As the Covered Wells stick records, federal wells came
with the teens, but as late as 1919 the Indians had 16,000 acres in
floodwater farming (Castetter, E.F., and W.H. Bell, 1942). The
land barren of living streams held few large aquifers. Three valleys
had underground water capable of pump rates of more than 200
gallons per minute. This was enough for stock raising. As late as
1949, only about 40 percent of Papaguería could be grazed; the
absence of water kept the cattle off much of the land (Kelly, W.H.,
1963).

A small reservation (70,000 acres) was created on the Santa
Cruz in 1874. The bulk of the Papago lands were set aside in
1916, nearly 3,000,000 acres. A federal school was started, a hos-
pital begun. In this absentminded manner the Papago came under
the care of the Bureau of Indian Affairs. The link between the
people and the resources of their land had not yet been broken.
The federal government expressed satisfaction with this unique
situation. The Board of Indian Commissioners stated their inten-
tions in 1919:

If the tribe is helpfully aided in education, water, health, and

One heard a drone, a hypnotic drone in the desert night.
Boys must run and run and run. Girls must grind corn.
Do not be idle, the voice would instruct.

O-otam potmaker. Smithsonian Institution Photo No. 2753.

Ball player. Smithsonian Institution Photo No. 2771-B.

Water carrier. Smithsonian Institution Photo No. 2752-C.

Burden. Smithsonian Institution Photos Nos. 2747-A, 2747-B,
2747-C, 2747-D, 2747-E.

livestock improvement, and subjected to the minimum of government jurisdiction, and that only of the most helpful kind, there is little question that, within a generation, the Papago Indians will be self-supporting citizens of the United States and Arizona, respected by their white neighbors and a most valuable national asset. (Underhill, R.M., 1939)

Federal plans would make a self-sufficient people self-sufficient.[16]

Something happened. Education divided the tribe between those who had seen the tractor and those who had not. Stimulation of the cattle industry resulted in the ruin of the rangelands. The wells, meant to permit escape from the motion of two-village life, cut the ground from underneath the ancient mutual dependence and sharing. A half century after the commissioners' optimistic forecast, the Papago are not respected by their white neighbors and are not self-supporting. They now have a groundwater problem, an overgrazing problem, and an economic problem. The society of abundance is gone.

After centuries of being a substitute for deer, cattle gained importance in the twentieth century. Between 1900 and 1925, 14 ranches were organized. By 1960, 12 more appeared (Bauer, R.W., 1971). Well drilling kept pace; the thirties found 56 wells, 18 dams, 41 masonry tanks, and 76 dirt tanks (Kelly, W.H., 1963). At the start of the First World War, the desert O-otam ran 30,000 to 50,000 beeves, and 8,000 to 10,000 horses. Five years later, estimates pegged the total at 30,000 cattle and 30,000 horses. The herds shrank to 27,000 cattle in 1939, and 18,000 horses. The year 1950 found only 13,000 cattle, and 7,000 horses. Today, cattle oscillate between 10,000 and 20,000 head; horses are greatly diminished in numbers (Bauer, R.W., 1971).

The herds are gone because disease destroyed the horses, and cattle destroyed the range. By 1960 about a third of the reservation, 1,250,000 acres, had lost three to nine inches of topsoil (ibid.). The Papago have fought all efforts to place limits on their stock numbers. The Bureau of Indian Affairs constantly criticizes them for their one wholehearted acceptance of a bureau program: cattle raising.

At first glance it appears that the Papagos took up white concepts of mining resources, and by leaving the society of balance and abundance, gutted their ancient lands. This first glance is wrong. The cow proved lethal to Papaguería because there was no control within the society of sharing to check it, and because there was no mechanism within this society for controlling those few Papagos who abandoned ancient ideas of generosity for personal greed. These two antithetical tendencies, group solidarity and individual enrichment, led to the same end. The traditional family raised cattle not for market but for kinsmen and needy O-otam. The land held in common became common grazing ground. Selling cattle for money was disdained. As late as 1968, a member of the Gu Achi Stockman's Association gave the following reasons for keeping as many cattle as possible: first, cows are banked as a source of money against unexpected demands for cash—a pilgrimage to Magdalena, for example; secondly, beeves are required for saint-day feasts, weddings, and ceremonies; finally, cattle are a commodity for trading with other Papagos for beans, and so forth (ibid.). The white superintendent of the reservation explained in the late sixties, "Their priorities are backwards" (ibid.). Those few Papago who saw the cow as a way into private enterprise (at most half a dozen families) exploited the reservation without restraint in search of immediate profit. The other O-otam did not try to stop them. By taking the path of greed and miserliness, these families were socially dead. They had violated the way of the desert people; they did not exist.

The circle is vicious. The ancient way was based on the careful use of renewable resources, and the rate of resource availability was dependent upon an undependable hydrologic budget. Change this budget and the way either collapses or grows grotesque. As the federal government took command of the Papago society, the people lost their ability to adjust to new resources and new sources of power. Wells were drilled without their consent, children were sent to schools against their protests, tribal structure was dictated to them in the name of representative democracy.[17] Cattle fitted into older patterns: work was communal, the old supervised the young, the productivity was easily shared, the land continued to support the people. The Papago seized the cow as a shield against outside encroachments on their society. The tactic failed because the cattle had grown independent of control by the desert. Veterinary medicine conquered disease; natural predators were killed or proved not up to the job of overcoming steers; wells and ponds kept expanding the amount of ground the cows could graze. Water, vegetation, and soil began to circulate at velocities that changed the face of the land. Papaguería became an example of eroded, overgrazed terrain, and the Papago cattle industry became the Papago cattle problem.

Papago experience in the twentieth century has seesawed between Indian attempts to hold on to the society of abundance and white efforts to prod the Papago into the American economy. The Americans have largely won. Floodwater farming, once the psychic key to sharing, is almost gone. It was described this way in a 1949 study: "Flood water farming should be discouraged. It is too risky, unproductive and erodes the soil . . . In view of the rather complete change Papagos have made from their old barter and gift economy to a cash economy integrated with the wider American economic system, the subsistence farm would appear to be outmoded" (Dobyns, H.F., 1949).

Cash labor off the reservation has replaced farming on it. Foraging is about dead; the 1949 study lamented this abandonment of wild foods. The modern diet, the report concluded, was not as nutritious (ibid.). There has been scattered resistance to change. Reduction of the cattle herds is resented. Well-drilling rigs sometimes lurched into situations "where the chief met the well drillers on the road with a rifle . . ." (ibid.). Various factions within the tribe cling to the past or plead for the future. The pace of change is slowed by remnants of the O-otam tradition of group harmony; opponents avoid confrontation and simply try to outlast each other.

What is left of the range supports three head of cattle per square mile per year. Two large mining companies are gearing up to exploit copper deposits. Some baskets, perhaps 3,000 per year, are sold. Stock sales bring in $1,000,000 a year. The average per capita income runs about $700. More than 10,000 O-otam claim the reservation as home. But the values cut into the Covered Wells stick are fugitives in Papaguería. "The livelihood of all Papagos living at San Xavier depends upon a cash economy. No one gathers, hunts or grows or raises his own sustenance. Subsistence depends entirely on one's ability to get cash . . ." (Fontana, B.L., 1960). The backward priorities are being straightened out.

This change is sometimes called progress and this is held to be good. This change is sometimes called the destruction of a natural way of life and this is held to be bad. Or, to avoid the good and the bad of it, the change is called acculturation. By any name, the change is crucial. The Papago, like many peoples on the planet, have moved from an economy based on the use of renewable resources to an economy based on the depletion of nonrenewable resources. In shifting from one economic base to another, they have altered their values, their families, their sense of duty and obligation. Groundwater exploitation has been central to this metamorphosis. The pumped water killed the two-village system and expanded the cattle industry. Electricity and gasoline furthered this move. No longer living off the land of Papaguería, the O-otam now number far more than that land can probably sustain. They have grown in ways parallel to their white neighbors.

The impact of groundwater development is to leave a culture of balanced resource use for one of imbalanced resource use. The mining of water buried in the earth encourages the mining of other resources. The O-otam have lost the Keeper of the Smoke, given up the call of the dream in the night, turned a deaf ear to the voice in the darkness. In lives bunkered by steel, gasoline, electricity, groundwater, and cash, they can ignore the desert for a while.

5. Conquest

The names come from other tongues. *California* springs from some Spanish dream of gold and pearls; *Colorado* simply means red; *Nevada* stands for snowcapped mountains. Pimas described a place having a little spring with *Arizona*. Navajos meant only land of the Utes: *Utah*. Conquistadors seeking El Dorado wrote *New Mexico* on the map. Shoshones said something with *Idaho*. *Montana* resulted from forcing the Spanish *montaña* to suggest mountains. The Great Plains are littered with Sioux ideas. The *Dakotas* stand for allies; *Kansas* indicates the South Wind people. Omahas expressed flat and shallow with *Nebraska*. Red people is the Choctaw notion buried in *Oklahoma*; *Texas* a Caddo way of saying, "Hello, friend." Wandering Delawares conveyed "upon a great plain" with *Wyoming*. The arid West is splattered with a babel of languages.

The richness of the words misleads; the color of western history has often overwhelmed the content. The boys who punched cows, the shaggy men who killed the beaver, the Indians who were fodder for cavalry, the miners who searched for a heart of gold, the individuals of exotic reflexes who could dispense small bits of lead at remarkable rates—these humans have had greater impact on modern minds than they had on the land. The grass-shorn rangelands, mine-pitted hillsides, and declining aquifers testify to capital from the eastern United States and from Europe seeking a profitable return. Now such activities are called development; a century ago it was simply destiny. "When Uncle Sam puts his hand to a task," explained one advocate of such policies, "we know it will be done . . . When he waves his hand toward the desert and says, 'Let there be water!' we know . . ." (Smythe, W.E., 1911).

The chore of concentrating and exploiting resources has gone on beneath the clatter of cowboys and Indians. It is the real stuff of western history, and to look at this process is to gain an understanding of arid lands in general. Because water is scarce, photosynthesis is limited. This means many resources are cycled at slow rates. Humans in the past century in the western United States have used them at fast rates. For decades beeves were shipped east by literally mining the grass and the topsoil. This resulted in devastation of the rangeland, and finally, in the Taylor Grazing Act. Today, management is supposed to keep cattle numbers in balance with the productivity of the vegetation. Clear cutting of the forests led to the same degradation and culminated in yet more federal laws to guarantee a rate of timbering the woods could sustain.

But all resource exploitation does not offer such an easy adjustment to human appetites. Minerals are a gigantic industry in the arid west. Regardless of the rate at which they are ripped from the earth, they are essentially nonrenewable. That is why mining towns end in collapse. The coal beds, oil pools, uranium deposits, and oil shale belts are also one-shot affairs. The energy industry moving into the region can be only a temporary and transitional kind of land use.[18] Water presents a special case. The rivers and other surface supplies are renewed by the rain, and hence, are permanently available. Americans seized this resource first. During the 1930s, 400 million dollars plugged the Colorado with dams and made a wild river a piece of plumbing. In varying degrees, this has been the pattern throughout the west. Since surface supplies were spoken for long ago, Americans have turned to groundwater. Arizona meets 60 percent of its water needs by pumping aquifers. Other states and basins approximate this dependence. The water is commonly pumped at rates faster than natural replenishment. This is called mining.

Groundwater is essentially nonrenewable in the arid west because the economies that exploit it cannot abide a low rate of use. By combusting nonrenewable coal and nonrenewable oil and nonrenewable natural gas, they have managed to lift nonrenewable water at incredible rates. By using water with abandon they can compete with more humid regions, where it is basically a free good. This extractive process, like the looting of ore deposits, soil, forests, and fuels, is the machinery behind the expressions "conquest of nature" and "the miracle of the deserts." Rip away the veneer of western history and this consumption of resources links the centuries.

The last cut

There seems to be no turning back. As the record of the Covered Wells stick ended in the 1930s, an anthropologist, Ruth Underhill, raced to capture the contents of traditional O-otam society before they vanished. She would ask people about old ways and old ideas. She carefully noted the changes being caused by federal programs, by the entrance of the cash economy among the people, and by the shift from old ways of using resources to new ways. One question flung at her by a Papago made a vivid impression. "Why can't the school teach us to make cake?" demanded the person. "We can buy eggs and milk. We want to make everything the white people make" (Underhill, R.M., n.d.).

The practices and appetites of western industrial societies spread very easily. The twentieth century has witnessed people after people leaving ancient traditions for these new habits with little thought as to the consequences and little control over the process. Scholarship on development has remained largely a futile literature; the movement from a solar-fired economy to a fossil-fueled economy seems to have a life of its own.

Among the Piman people stick cutting ceased in the thirties. The water tanks, roads, electric lines, wells, and new jobs produced a swirl of events that gouges and nicks could not order and capture. One stick belonged to a medicine man. He was murdered and his wand vanished. Another was tossed in the trash by a widow. There is a rumor that at least one stick lingers in Papago hands; it is not known if anyone can read it. The records slashed in saguaro are gone because the messages once left in the wood are no longer enough. The mind that could ignore a war and remember a big snow is dead. The desert people, white and O-otam, have joined together in a powerful occupance of the land. They have gathered around pumps.

Part Two

Down to this time our apparently wasteful culture has, as I have sought to show, been the true economy of the national strength; our apparent abuse of the capital fund of the country has, in fact, effected the highest possible improvement of the public patrimony. Thirty-eight noble states, in an indissoluble union, are the ample justification of this policy. Their schoolhouses, and churches, their shops and factories, their roads and bridges, their railways and warehouses, are the fruits of the characteristic American agriculture of the past.

*—General Francis A. Walker,
Preface to Volume III of the
Census of 1880*

When I was little any time my daddy walked up to a windmill he'd take off his hat and grin up at that fan going around. Just grin like an idiot.

—Recollection of a plainsman

1. Pump

The days and the nights have ceased to matter on the High Plains of west Texas. After a century of trial and error and error and error, this vast tableland hosts a petroleum industry, one million humans, and huge irrigated farms. "People from the eastern part of [Texas]," a reporter explained, "who might drive through Crosby County at night are often amazed by the flashlights and electric lanterns darting over the fields and highways. They usually stop at the nearest service station and ask, 'What's going on?' Often they get fantastic answers ..." (Green, D.E., 1973).

The curious travelers have stumbled upon an army directing water across the High Plains. Pickups bounce down country roads so that this legion of people can move irrigation pipes at two, three, four in the morning. Flashlights bob in the summer night for men linking ditch to furrow so that crops can drink. This work goes on in a din of sound as pumps suck water to the surface of the plains. Man, his livestock, and his plants satisfy their biological appetites in an environment of machines, chemicals, and pumps. By day one can see huge metal creatures crawling across the fields. These high-towered, tandem-wheeled behemoths are mechanical irrigators, a substitute for the humans shifting pipe through the night. They represent one more American effort to remove human labor and natural delivery of materials from agriculture.

By a heavy investment of resources and of energy, the High Plains of Texas have emerged a major source of food and fiber. Groundwater pumping has made rain an intruder in the water schedule. Chemicals have altered manure from a fertilizer to a pollutant. Insecticides have cast all insects as targets. Herbicides have divided the botanical world into two categories: crops and weeds. The very best of modern American agriculture has been brought to the High Plains. The humans of the United States have been described as a "people of plenty" (Potter, D.M., 1954). This flat Texas land is as close to the core of the plenty as one is likely to get.

The planet harbors two distinct groundwater problems. In underdeveloped countries the challenge lies in exploiting the aqui-

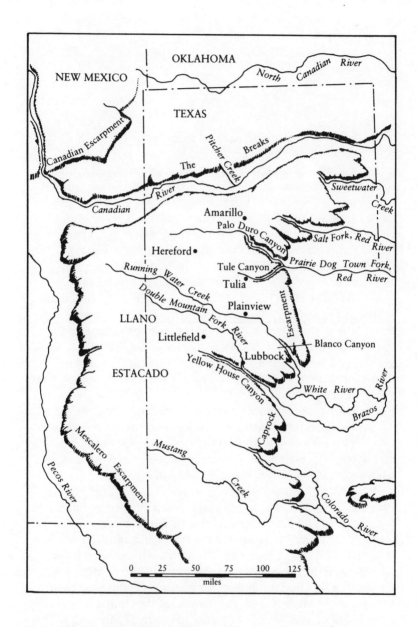

MAP 4. Texas High Plains: Physical. Based on the map by J. L. Rogers in *Land of the Underground Rain: Irrigation on the Texas High Plains, 1910–1970*, Donald E. Green, Austin: University of Texas Press, copyright © 1973.

fers. In developed countries studies focus on the consequences of such exploitation. West Texas has experienced both problems, and fashioned answers to each. It was once a hungry place where the greatest concern was getting at resources, not running out of them. It is now a rich place, and the various pumps of resources, whether oil rigs, livestock, plants, or water wells, wrestle with the reality of using vital elements faster than they are replaced. The inhabitants—human, botanical, and herbivore—face being ousted from the land. The High Plains rest on a tub of water called the Ogallala aquifer. The water is stored in sand and gravel laid down during the late Miocene and Pliocene. This debris is the eroded material stripped from the Rocky Mountains and dropped in ancient valleys. Later warping left the alluvium vulnerable to severe erosion. A vast tableland was created fronting the mountains: the High Plains. The Texas portion of this geology is cut by the Canadian River near Amarillo in the Panhandle. Called the Llano Estacado, or Staked Plains, by the Spaniards, this chunk of Texas runs about 35,000 square miles, and in 1968 witnessed 73 percent of the state's groundwater withdrawals. Put another way, Texans consumed twice as much water from the Ogallala that year as from all the surface storage in the state. Thickness of the waterbearing layering runs 140 feet on the average (Conselmen, F.H., 1970).

Starting at about 4,000 feet above sea level in the north Panhandle, the plains sink to 1,200 feet in the south near the Pecos Valley. Rock escarpments on the east and west separate the flat tract visually and deny the possibility of lateral recharge to the Ogallala. Recharge comes only from the rain, and the rain varies from 12 inches in the south to 20 in the north. The plains are too flat for lakes, and the evaporation rate too fierce for standing water. Most of the precipitation is lost to the atmosphere. What little sinks into the soil recharges the Ogallala. The rate is estimated at a twentieth, a quarter, or a half inch per year. Triassic red beds of clay beneath the water-bearing sand and gravel seal the tub. Seventy-five years ago, the aquifer possibly held 300 or 400 million acre feet of water under Texas. This buried lake made the economy of modern west Texas; pumping at rates greater than the half inch of recharge per year made the groundwater problem of west Texas (Graves, J., et al., 1971; Green, D.E., 1973).

By 1959 irrigated agriculture supported an economy of $330 million on the High Plains (Green, D.E., 1973). Ten years later, this had grown to $1.1 billion, two-thirds of which was cotton and grain sorghum. Cattle boomed in the sixties, and 1,792,000 steers generated the rest of this billion dollar bonanza by 1969 (Osborn, J.E., and W.C. McCray, 1972). Farms were big. An insurance appraiser in 1967 felt that a section (640 acres) of irrigated fields was a minimum unit (Green, D.E., 1973). High Plains agriculturalists agreed. The plains as a whole averaged around 3,500 acres per operation in the sixties, with the heavily irrigated south plains running close to 1,000 acres (Osborn, J.E., and W.C. McCray, 1972). In the mid-twentieth century, the Ogallala hosts a billion dollar farming economy and a million people (ibid.).

In the late sixties the tableland had 5,500,000 acres under irrigation and 45,000 wells delivering the water (Green, D.E., 1973). Feedlot cattle grew during that decade, and with 1970, close to three million head lived in giant lots. These captive herds required 650,000 irrigated acres for food and prompted 80,000 acres to be put into irrigated pasture. In some instances, computers were employed to coordinate feeding the beasts. Eleven High Plains slaughterhouses greeted the fattened steers (ibid.). Exploitation of gas and oil since the 1930s has provided cheap energy for this movement of water and cattle.

Pick up a pen and a scrap of paper and the thing draws itself.[1] First, sketch an enormous tract of flat land underlain by an enormous amount of buried water. Add energy; tractors to level fields; pumps to lift water; chemicals to fertilize the earth, slaughter insects, and destroy weeds. Pour gasoline into engines and monster irrigators lumber across fields; huge machines harvest cotton, wheat, and sorghum. Build steel pens and granaries and plug in a computer, and cattle stuff themselves toward beefsteak. Pump fossil water and fossil fuels as fast as possible and the High Plains add up to a billion dollar economy. This velocity of resource use has sent humans into the night bobbing flashlights.

If there were an endless supply of water and concentrated fuels, the activity on the High Plains would also be endless. But this food factory rests on storages of water, gas, and oil, and these storages are running out. The Ogallala, named after a broken Dakota band, is being broken. By the 1980s water declines should make serious inroads in irrigated agriculture; thirty or forty years hence this commerce of pumped water should be over. The humans of the High Plains will be staring down tens of thousands of dry holes.

Like the Sonoran desert with its waves of O-otam, Spaniard, Mexican, and American, the High Plains have witnessed successive cultures seeking fortune in the land's resources. But the events have happened more swiftly on this tableland. The technology which many now look to for a fix has helped invent the booming economy of the High Plains since the end of World War II. Most of the water has been pumped in the last thirty years. Whites have only been on the plains for a century. Their brief hundred years on the land has been a struggle to fashion machines and find crops capable of plundering the earth and the aquifer beneath it. Behind the whites lies a group who never farmed, pumped, or herded. The journey from a culture fostered by renewable resources to one dependent on mining resources has taken only ten decades, and most of the change has occurred in a mere three. On the Llano Estacado the emergence of the groundwater problem is compact. Significant human exploitation of the region starts with a people known by the Ute word for enemy: Comanches.

Grass

They never planted a seed, plowed a furrow, herded a cow, or dug a well. The Comanches spilled out of the Great Basin unto the plains before Europeans touched North America. They straggled into the sea of grass with their few bits of property carried by their women or dragged by their dogs. When the horse fell their way the naming was obvious: God-dog. Called enemy (Comanche) by the rest of humanity, they described themselves as Nermeruh, The True Human Beings (Fehrenbach, T.R., 1974).[2] Their domain reached north into Kansas along the Arkansas River, then, running south, sliced Oklahoma in half and continued to the Balcones Escarpment above San Antonio. From there it swung west to the Rio Grande, and then reached north into eastern Colorado and the fork of the Arkansas. The High Plains formed the heart of their world. Living by hunting and gathering, the Staked Plains storages of fossil water and fossil fuels were of no account to them. Like most aboriginal groups, they suggest the potential of a society based on the consumption of renewable resources. The Nermeruh also offer one further insight into the process now called development. By acquiring the horse, the group experienced a sudden infusion of energy and enjoyed a boom comparable to that of modern Texans looting the plains water and petroleum reserves.

MAP 5. Comanchería. By permission from *Comanches: The Destruction of a People*, T. R. Fehrenbach, New York: Alfred A. Knopf, Inc., copyright © 1974.

They went gentle on the land.

Noah Smithwick, visiting on behalf of the United States government in 1837, got a lecture on land ethics from a Nermeruh chief:

> We set our lodges in these groves and swung our children from these boughs from time immemorial. When the game beats away from us, we pull down our lodges and move away, leaving no trace to frighten it, and in a while it comes back. But the white man comes and cuts down the trees, building houses and fences, and the buffalo get frightened and leave and never come back and the Indians are left to starve, or if we follow the game, we trespass on the hunting grounds of other tribes and war ensues. (Fehrenbach, T.R., 1974)

Americans required a full four decades to budge the tribesmen from this philosophy.

They went hard on strangers.

An early Spanish account noted their treatment of captives taken from the neighboring Tonkawas. First, the Nermeruh staked out their guests and generously applied fire to the hands and feet. When destroyed nerve endings blocked sensation, they would whack off a chunk of the limb and repeat the applications of fire. This process continued until night fell and boredom set in. Then, the Tonkawas were scalped (a ritual form of insult on the plains), and the Comanches, looking forward to a sound sleep, ripped out their tongues. The mute victims were finally tossed onto the fire with heaps of coals on their bellies and scrota. The Comanches dozed around the blaze until morning. Such instances could be almost endlessly repeated (ibid.).

The True Human Beings lived a world of motion. Their vast holdings on the plains had no real center. Meat was the buffalo, and humans followed the herds on horseback. Housing was portable, the tipi. Produce came from fifty-two wild plants (almost half were a source of medicine and materials). Fruit was had from persimmons, mulberries, wild plums, grapes, currants, juniper berries, hackberries, prickly pears, and sumac. Root plants utilized included Indian potatoes, onions, radishes, Jerusalem artichokes, and sego lilies. A few dried fruitcakes were stored. Trade brought the nomads corn, pumpkin seeds, and tobacco. Honey was a periodic delight; beer and wine were not made. A favorite dish was crushed buffalo marrow and mashed mesquite beans. Pemmican, the dried staple of the plains, constituted the basic ration (Wallace, E., and E.A. Hoebel, 1952).

Motion and tonnage did not mix. Faced with the erratic weather of the semi-arid plains, the Nermeruh opted for pursuit of the buffalo over massive storage of the vegetation. Often hungry, they amazed early Europeans with their capacity for gluttony when a feast came their way. The bison was the Comanche granary, warehouse, and insurance against starvation. This herbivore concentrated the expanse of grass in one fat body; in winter the bounty of summer still existed in buffalo flesh. Just as oil pools represent concentrations of solar energy, so do buffalo; only the rate of replenishment and the degree of concentration vary.

The constant march through the grass was hard. Sometimes water holes turned up dry, and weary warriors with swollen tongues would push the group on, threatening death to any who stole the little water left for the children in the buffalo paunch canteens. Nermeruh women had many miscarriages, and frequent barrenness (Fehrenbach, T.R., 1974). The pain came from the motion and the motion came from the buffalo. This shaggy beast evened out the High Plains oscillations of weather and succession of seasons.[3]

Plains life was dependent upon buffalo, but made possible by the horse. "Steam, electricity, and gasoline have wrought no greater changes in our culture," argued historian Walter Prescott Webb, "than did horses in the culture of the Plains Indians" (Wallace, E., and E.A., Hoebel, 1952). Long vanished from the North American continent, the horse returned in European hands to dazzle humans who could control no beast larger than a dog. It slipped slowly from Spanish control.

As late as 1650 Spanish records make no mention of any Indians north of Sonora possessing horses. With Kino's penetration of Pimería Alta in the late seventeenth century, Apaches seized the new beast for transportation and food. But expeditions of Spaniards onto the plains fronting the Rockies continued to find Indians on foot up to 1675. The great Pueblo revolt of 1680 saw large herds escape the murdered Europeans of New Mexico. In 1681, the Mendoza-Lopez *entrada* scouted west Texas. Mounted Apaches attacked along the Rio Pecos. It had begun (Fehrenbach, T.R., 1974).

Before the horse, humans could not reap the riches of the grasslands. Man on foot could not keep up with the buffalo. Small nomadic groups, like the Comanche, nipped at the herds and barely survived. Major populations on the plains sought shelter in the river valleys tilling a little ground and feasting once in a while on a bison (Holder, P., 1970). The horse gave access to the buffalo and enormously increased the amount of energy available to the people living on the plains.[4]

This new access to resources was as radical for the Indian as groundwater and petroleum would be for the Americans. Solar energy was concentrated in the grass, and buffalo by consuming this caloric bonanza made the resource digestible and feasible for humans. The horse acted as a pump. Mounted and armed with bows and lances, the Indians sent the flesh of bison pouring through their societies. The plains people became legendary horsemen slashing through black clots of buffalo. Their practices and appetites made no appreciable dent on the enormous herds. The culture of motion, tipis, and buffalo could have continued forever.

The Nermeruh living close by Mexico became the main horse jobbers for the entire Great Plains. Their Shoshone dialect served as the language of commerce in the grasslands. By the nineteenth century, one Comanche group, the Antelope band, contained 2,000 humans and 15,000 horses. A warrior might possess a remuda of 250 head, a chief 1,500. Farther north horse numbers declined with distance. The entire Pawnee tribe owned 1,400, the Osage and Omaha 1,200 apiece. A Dakota leader would be exceptionally rich with fifty mounts (Wallace, E., and E.A. Hoebel, 1952; Fehrenbach, T.R., 1974).

Any man with a horse and a weapon had equal access to the key resource: buffalo. In modern Texas any man with a well has equal access to the Ogallala aquifer. This democratic fact led in both instances to an anarchy of resource use. For the plains Indians, the result was a very loose tribal structure, individual assertion of whim and right, and personal violence (Holder, P., 1970). Bands moved at will, and warriors left bands impulsively. The discipline and group solidarity of the O-otam seemed unnecessary in the grasslands. The True Human Beings ruled their empire for three hundred years without any real tribal structure, without a head chief, without ever meeting together. The first and last joint assembly of the Comanches came in 1874 at the first and last Comanche Sun Dance. This singular event was a futile effort to stave off white pressure on their lands (Wallace, E., and E.A. Hoebel, 1952).

For decades following 1680 the Spaniards did not grasp what was happening on the plains. The eastern Apaches suddenly moved west into New Mexico, mauled and broken. Still, the Spaniards blamed the new eruption of violence on the Apaches. The Nermeruh invented Comanchería by stealing Spanish horses, copying Spanish bridles, saddles, and stirrups. They carried lances just like the Spaniards. During the long wars with the Moors, the Spaniards had learned to mount horses from the right. All Comanches mounted from the right (Fehrenbach, T.R., 1974). While this transmission of technology and culture was taking place, the Europeans had no word for the Nermeruh and did not realize they existed.

The True Human Beings ended this misunderstanding. Having gained access to the riches of the grass, they put their surplus energy to work. The plains tribes in general, and the Comanches in particular, turned to raiding. The people with an abundance of horses took to stealing more horses. For a century and a half, the Nermeruh robbed, murdered, and raped European settlers with regularity. The horses returned from the grasslands with homocide as a cargo. The range of this violence was immense. One Kiowa band based in Kansas was said to have struck south as far as Yucatán, returning with strange talk of parrots and monkeys (ibid.). The Comanches regularly pillaged within five hundred miles of Mexico City. For a thousand miles south of the High Plains the full lunar phase came to be called the Comanche moon. Raiders rode to Chihuahua, Zacatecas, Coahuila, San Luis Potosí, Nuevo León, Durango (Wallace, E., and E.A. Hoebel, 1952).

"For days together," reported an 1846 traveler in Mexico, "I traversed a country completely deserted . . . passing through villages untrodden for years by the foot of man" (Ruxton, G.F.A., 1847). The Spaniards countered this threat with a few expeditions onto the plains; they were slaughtered by whirling horsemen wearing buffalo heads. The Mexicans' only weapon was disease; Comanchería was wracked with cholera, smallpox, and venereal maladies (Fehrenbach, T.R., 1974). Americans collided with this buffalo culture wall in the nineteenth century. Prior to the Civil War the Texas frontier averaged one hundred humans per thousand square miles. This line of settlers crawled westward at a cost of seventeen lives per mile of advance (ibid.).

The raids resulted in death, rape, horse stealing, and captives.

Humans taken alive were raped if women, tortured if men, and once in a while sold back for money. When Americans in the 1870s offered $100 for each victim returned, the Nermeruh made a little industry of it. Comanche shopping lists were varied. One raid into Texas in the 1840s left Daniel Boone's granddaughter dead, a coastal town sacked, and Indian saddlebags bulging with law books—esteemed for rolling cigarettes (ibid.). The homocidal expeditions left trails still visible, the Comanche traces, radiating from the High Plains. In the early twentieth century, J. Frank Dobie found Mexican parents still terrorizing their children with threats of Comanche murder when the moon grew full (Dobie, J.F.,1935). Such were the consequences of the first human efforts to succeed in pumping resources from the grass of the plains.

Under Comanche tutoring, the American army was honed to modern warfare just prior to the Civil War. At first the troopers rode heavy horses and were armed with single-shot muskets. Nermeruh cut them to ribbons with fast steeds and fourteen-foot lances. Samuel Colt was saved from bankruptcy because his new six-shot revolver proved wonderful for plains combat. Faced with the plains warriors, the army slowly let go codes of war and turned with zeal to genocide. Indians were killed without regard to age or sex, villages burned, captured horses sold or shot. Campaigns went on winter and summer. After visiting troops stationed against the Nermeruh, General Philip Sheridan remarked that, given a choice, he would rent out Texas and live in hell. The final tactic was to end Indian access to the energy of the grass. When the Texas legislature considered protecting the buffalo as a resource, Sheridan lectured them:

> White hunters assist the advance of civilization by destroying the Indians' commissary; and it is a well-known fact that an army losing its base of supplies is placed at a great disadvantage. Send them powder and lead, if you will, but for the sake of lasting peace, let them kill, skin and sell until the buffalos are exterminated. Then your prairies can be covered with speckled cattle and the festive cowboy who follows the hunter as a second forerunner of advanced civilization. (Fehrenbach, T.R., 1974)[5]

Due to a new process in the American leather industry, buffalo hides went for $3.75 apiece by 1873. In two years, 7,000,000

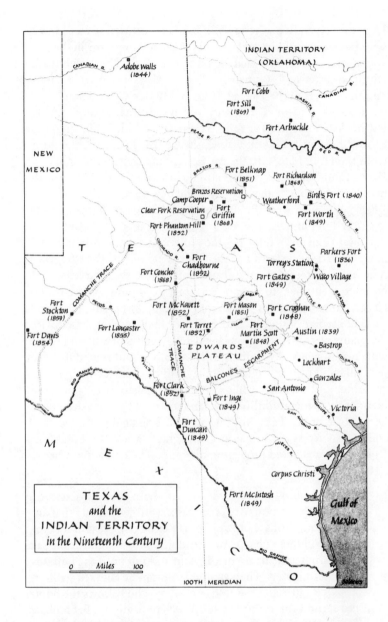

MAP 6. Texas and the Indian Territory in the Nineteenth Century.
By permission from *Comanches: The Destruction of a People*,
T. R. Fehrenbach, New York: Alfred A. Knopf, Inc., copyright
© 1974.

pounds of bison tongues were shipped from Dodge City. Between 1868 and 1881, the state of Kansas exported the bones of 31,000,000 buffalo for fertilizer. By 1881, there were no more shipments of rawhides, tongues, or bones. The buffalo had been exterminated. The Nermeruh had always left the heart in their kills to ensure the health of the herds. They could not believe the new reality (Fehrenbach, T.R., 1974).

Their last effort against the new order came at Adobe Walls in Oklahoma territory. Here, in 1874, the bands united for their first Sun Dance. By chance, they stumbled on a party of buffalo hunters. The two cultures, the two energy systems, collided with a sort of poetry. The Indians believed the dance made them immune to bullets. The hunters were armed with .50 caliber Sharps. A lucky shot cut down a warrior holding a council. The range: 1,500 yards. That ended it. Nermeruh will to resist was shattered by the crack of a rifle (ibid.).

When it was over, American soldiers rode with scalps dangling from their saddles, and only one Comanche in twelve had survived the previous quarter century. Pushed onto reservations the nomads stumbled toward sedentary life. Whites planted crops for them; the Indians looted the patches of green watermelons and grew violently ill. They decided farming was "bad medicine." Raised on fresh meat, they hated salt pork. Cornmeal was fed to the horses. Soap was dismissed out of hand: it tasted vile. When cattle were issued, the Nermeruh hunted them like buffalo. Of the ten, twenty, or thirty thousand Nermeruh of their heyday, maybe 1,500 survived (ibid.).

The Comanches, and other buffalo people, have endured as the dominant aboriginal image in the American imagination. Their independence and mobility spoke to deep hungers in their conquerors. Their impact on the plains themselves was so slight as to leave almost no traces. A few trails, a few campsites. This inability to make much of a dent in the storages of the grasslands is their relevance. As one chief, Ten Bears, explained, "We wish only to wander on the prairie until we die . . ." (Fehrenbach, T.R., 1974). Motion, tipis, men on horseback: this is the only known model of a steady-state economy on the High Plains. The hardship of their lives offers an insight into the price humans pay for a balanced use of resources. The explosive barbarism exhibited by plains societies when they gained the horse is a haunting example of human reactions to sudden infusions of energy.[6]

Wind

There were cowboys but no Indians. The void left by the extermi-
nation of the plains societies was first filled by cattlemen. Huge,
unfenced spreads seized the grass once devoured by the vanished
buffalo. Between 1875 and 1883 the Texas frontier marched farther westward than it had managed in the previous four decades
(Fehrenbach, T.R., 1974). A legion of farmers followed the cow-
men into the Dakotas, Nebraska, Kansas, eastern Colorado, Oklahoma, and west Texas. Blizzards in 1886–1887 destroyed the cattle
herds. Oscillations of weather broke successive waves of agriculturalists. The cow proved no substitute for the bison, the cultivated
crops no match for the native grasses. During the last quarter of
the nineteenth century, Americans tried to move their practices of
livestock raising and their habits of farming onto the arid and
semi-arid grasslands. They attempted to live within the solar cir-
culation of resources that had sufficed for the plains tribes. They
failed (Smith, H.N., 1950; Webb, W.P., 1931).
   One human explained it this way in 1906:

> . . . There is a stretch of country whose history is pregnant
> with greater promise than perhaps any other equal expanse
> of territory within the confines of the Western hemisphere.
> Following the times of occasional rain, this line of social
> advance rose and fell with rain and drought, like a mighty
> tide beating against the tremendous wall . . . And every such
> wave left behind it a mass of human wreckage in the shape
> of broken fortunes, deserted farms, and ruined homes.
> (Webb, W.P., 1931)

The Great Plains devoured the money, wit, and muscle of a generation and bred a venom that entered American politics as popu-
lism (Hicks, J.D., 1955). These humans denied the aridity of the
plains and claimed rain followed the plow. When it did not, and
drought hit, they would briefly toy with the notion of irrigation:
"Tell the people of Nebraska that we are going to make homes for
millions of men; that in these homes irrigation shall guarantee
industrial independence . . . Nebraska's best days are before her"
(Carlson, M.E., 1968). Such were the thoughts of irrigation
crusader William E. Smythe in 1893. When the rains returned,
humans on the plains would say things like this: "We have passed

from the drouth period and have entered an era of old time moisture supply. We will now stop talking about irrigation" (Green,
D.E., 1973). Such was the advice of a South Dakota newspaper in
1896.

Plains agriculture became a search for a way to escape weather.
When words failed in this effort, interest increased in methods to
pump underground water. The first successful device was the
windmill. "The windmill," historian Walter Prescott Webb believed, "was like a flag marking the spot where a small victory had
been won in the fight for water in an arid land" (Sageser, A.B.,
1967). For the first time man reached into the Ogallala. The
aquifer's existence had been known since 1854, when a Swiss
geologist on the High Plains discovered that groundwater "on the
Llano Estacado . . . may be found everywhere" (Green, D.E.,
1973). By 1900 cattle ranches had hundreds of windmills for
watering stock.

The early mills were folk inventions. They appeared without
fanfare. An 1890 engineering text bemoaned the fact that "so
many otherwise well informed and observant men" did not realize
that thousands of the devices cluttered the plains (Wolff, A.R.,
1890). Built out of junk, they cost as little as a buck and a half
(Sageser, A.B., 1967). The U.S. Geological Survey observed in
1897 that "it is surprising to note how little definite information
there is in regard to the efficiency of windmills . . ." (Murphy,
E.C., 1897). States like Nebraska paid for surveys of the contrap-
tions (Barbour, E.H., 1899).

Wind, like sunlight, was a dilute form of energy. On the plains
it averaged 10 to 15 miles an hour (although it could and did die
for days at a time). A windmill with a 25-foot wheel in a 16-mile-
an-hour wind developed about 1.34 useful horsepower (Wolff,
A.R., 1890). An 8½-foot wheel in the same breeze captured 0.04
horsepower. The monster 25 footer could lift about 37 gallons a
minute. Maximum depth for commercial purposes was 70 to 80
feet (ibid.). This was enough water to satisfy a cow, or a garden,
but was not enough for real crops. Like the Comanche lance against
the buffalo, the windmill against the aquifers could not make a
real dent in the resource. The windmill stood between Americans
and the weather, and it was not enough.

Plains droughts drove some humans to the rim of magic. In
1871 Edward Powers, an engineer, wrote a book that pointed out
the correlation between big battles and big rains. Congress in the

nineties appropriated $19,000 to test this hypothesis. General R.H.
Dyrenforth supervised a cannonade at Midland, Texas. The rains
stayed away. In the early twentieth century, C.W. Post, the cereal
king from Battle Creek, Michigan, tested the same idea on the
High Plains using dynamite. He also failed (Eaves, C.D., and C.A.
Hutchinson, 1952). These campaigns, along with throngs of ingenious rainmakers and torrents of tent-meeting prayers, suggest
the despair of Americans trapped in the oscillations of High Plains
weather. Much of the High Plains had as much rain per year as
Copenhagen and were just a few inches shy of London. The difference between the water-rich Europeans and the drought-crazed
Americans lay in rain spacing and dependability (ibid.).

Eighteen thousand wagons left Nebraska for Iowa in 1891;
western Kansas lost half its population between 1888 and 1892.
Such small towns as Plainview and Amarillo clung desperately to
the High Plains as the nineteenth century waned. By 1890, perhaps 6,900 Americans endured where 10,000 to 30,000 Nermeruh had thrived. Local boosters spawned fanciful publicity:

> Come to the Panhandle for cheap lands; come for rich pro-
> ductive soil; come for health; come for seasonable summers
> and balmy winters; come and raise cereals, fruits, vegetables,
> sorghum, grains, grasses and forage; come and raise cattle . . .
> There no longer is another such country as the Panhandle
> awaiting development . . . (Green, D.E., 1973)

Those who came often reaped the whirlwind of dust storms and
bankruptcy. The aridity denied in the public press was duly ac-
knowledged in the farm foreclosures of the county courthouse.

Frederick H. Newell put exploitation of the plains in perspective
in 1902:

> The great problem, then, of obtaining water is that of pump-
> ing it at a cost so low that this operation can be performed
> with profit . . . the question is not simply to lift the water.
> *It must be lifted in large quantities, and, more than this, the*
> *cost of so doing must be extremely low—so low that it shall*
> *bear but a small proportion to the value of the crops produced*
> [italics mine]. (Ibid.)

The point is almost too simple to be readily understood. Americans
are presently staring into economic ruin on the High Plains because water is running out. Americans seventy-five years ago faced

economic ruin because they could not get at the groundwater. Cheap energy permitted humans to flee one ruin for another. Constant tinkering led to something more powerful than the feeble, dilute, renewable wind.[7]

Fossil

People on the plains needed a pump, a crop, a fuel, and money. Between 1900 and the end of World War II these pieces fell into place, and the groundwater problem emerged from the depths of the grasslands. The process unfolded without a plan and contrary to human expectations. It has been said that planning is the empty ritual of American society; efforts to exploit the Ogallala suggest why this is so. Groundwater consumption on the High Plains was minimal during the first three decades of the twentieth century. Real estate speculators tried to sell tracts as irrigated farms, but interest waned during periods of rain and water-delivery systems were unappealing. Decades of weather finally overcame resistance to artificial watering, and decades of improvements finally fashioned a suitable pump. The victim of these years of meteorological education and mechanical advance was the sodbuster, a man who spent his time, money, and family (from Canada to the High Plains) trying to make a go of farming the grasslands.

He was a creature of the banks and merchants, not the imaginary yeoman wearing homespun. As Lena Murphy explained at the 1913 meeting of the Texas State Farmers' Institute, his clothes were woven in the East, his bucket came from Missouri, his food often as not was bought at the store, his "meal was cooked on a St. Louis stove, with wood chopped with a St. Louis axe and hauled on an Indiana wagon" (Murphy, L., 1914). The only local thing on a man's place, she concluded, was the howl of his Texas dog. This debt-ridden citizen living in a dirt dugout sought not subsistence but the staples and productivity of commercial agriculture. For more than thirty years he and the High Plains lurched toward this goal.

The successor to the windmill was slapped together in a helter-skelter manner by people far from west Texas. Mahlon E. Layne of the northern Great Plains was in Texas for the oil boom of 1901. A well driller from the Dakotas, he had devised a technique for sinking large boreholes in order to prevent clogging. He brought this know-how to the oil fields and teamed up with a young engineer named O.P. Woodburn. Freak circumstance lured him into irrigated agriculture. The Republicans had just erected a new tariff wall which greatly stimulated the Louisiana and east Texas rice industry at a time when it was enduring subnormal rainfall. Pumped water was the answer to this drought, and the nature of the newly emerged centrifugal pump demanded large pits. Layne prospered drilling the holes and soon began modifying the pump.

The centrifugal pump had been around since 1754, but English inventors had greatly improved its performance by 1885. The device could lift far more water than windmills, but it could only function successfully within twenty feet of the water table. Farmers in eastern Colorado, Kansas, Arizona, and California had been toying with it. Expensive to buy and expensive to run, the pump promised to pay for itself in the high-profit rice industry.

To get within twenty feet of the water table, the pump often had to be sunk deep in the earth. "Imagine," Woodburn explained, "getting down into the pit to oil the pump with the mess of rope running at the velocity of the outside diameter of the 54 inch flywheel with six or eight 50 [pound] weights dancing on the tightener above your head. BAD DREAMS" (Green, D.E., 1973). By 1902 Layne had simplified lubrication enough to end the bad dreams. Within seven years he and his partner had peddled five hundred such pumps to rice growers at $500 apiece (without powerplant or well). This crude device still lacked a cheap driving power.

Steam was often the choice in the early installations, but such engines demanded a permanent crew and entailed a high operating expense. The budding electric utility industry sought the business, but this source required stringing lines to the fields (*Electrical World*, 1910; Adams, A.D., 1910; Reynold, Jr., E.C., 1910; Williams, C.H., 1911). Interest focused on a toy device developed in Europe in the mid-nineteenth century: the internal combustion engine. An English firm had built a low-compression, oil-burning version in the 1890s; by the early twentieth century Americans were manufacturing this design. The single cylinder ranged from five to seventy horsepower. Hard to start and cursed with a low efficiency, the engine ran pop-pop-pop. A seventy horsepower model cost $1,540 in 1913 (plus well drilling and pump), but this was in a Texas where oil sold for 3 to 7 cents a gallon (Green, D.E., 1973).

The engine was connected to the new pump by a belt, and this fact made the machine a hard mistress. In the summer of 1914 Roland Lloyd cut a chunk from his belt on June 29 to compensate for stretch in a heat wave. He sewed the piece back on July 21, cut again on July 30, and yet more on August 1. Because of the belt, the quirks of the pump, and the novelty of the engines, farmers sat with the machines when they were working, and never ran them at night. This automatically halved the maximum potential of water mining (ibid.). Granting these limitations, the import of the new technology was clearly understood on the High Plains by 1914. "The centrifugal pump," argued Zenas E. Black, "has lifted the shallow water portions of the Texas plains from bondage to the erratic cloud. In this work it has been assisted by the crude oil and distillate burning engine. The perfection of the above agencies has been the greatest boon that inventors have given the world during the last ten years" (1914).[8]

The great boon, midwifed by some engineers in the rice fields out of long known and unperfected technology, had scant success on the High Plains. Periodic wet seasons gave farmers the hope that they could avoid the $4,000 investment such an installation represented. Depressed farm product prices in the twenties also took the bloom off this wonder. By 1920 the four most heavily developed High Plains counties had only 187 irrigation wells, and 7,384 irrigated acres by 1930. One 74-acre irrigated tract returned its owner only $196.53 in 1924 (Green, D.E., 1973).

On April 14, 1935, Woody Guthrie killed time in Pampa, Texas, by writing a song called "So Long, It's Been Good to Know You." The occasion was a black duster which had shut down the plains. Successive dust storms ripped the region in the early thirties and made the Great Depression more savage for the plainsmen. Mass migration pouring from the grasslands picked John Steinbeck's grapes of wrath. Seeking a concluding image for his thirties trilogy, *U.S.A.*, John Dos Passos hit upon the idea of the vagabond. The arid barrier that had broken humans in the 1880s was back. Once again, the Great American Desert had said hello.

The economic and environmental ruin of the plains produced a remarkable 1937 report, *The Future of the Great Plains*, penned by a presidential committee for Franklin D. Roosevelt. "The people were energetic and courageous, and they loved their land," the joint authors believed. "Yet they were increasingly less secure on it" (Great Plains Committee, 1937). This sorry end had come to pass because Americans were in the wrong place doing the wrong things. Speaking of the plains in general, the committee argued that humans had plowed land better left to grass, had raised crops better cultivated elsewhere, and had tapped groundwater at rates wasteful and suicidal. "Where the rate of withdrawal from the underground reservoir is greater than the rate at which recharge takes place," explained the authors, "the falling water table causes the lift and cost of pumping to mount constantly" (ibid.). Economies based on rapid use of exhaustible resources rapidly collapse. The plains farmer was described as the victim of myths: That Man Conquers Nature; That Natural Resources Are Inexhaustible; That Habitual Practices Are Best; That What Is Good for the Individual Is Good for Everybody; That Expanding Markets Will Continue Indefinitely; and so forth (ibid.).

These mistaken notions were rooted in the size of the American empire. The country was so big nobody believed citizens of the republic could run out of country. The committee noted that the Secretary of the Treasury in 1827 had figured it would take "ages to come" to settle the West. Senator Thomas Benton of Missouri had estimated a year later that Americans needed "about 2,000 years more to complete the sales [of public lands] to the head of the Mississippi and to the foot of the Rocky Mountains" (ibid.). They were wrong, the report argued; in 1937, the search for riches in the arid grasslands was over.[9]

Irrigation, the committee declared, "at best can cause only minor changes in the economic life of the Great Plains." The region's aquifers had very little recharge: how can a growth economy be based on a resource that shrinks? "Increased drafts upon groundwater for irrigation," they concluded, ". . . are practicable only in a few favorable areas" (ibid.). American ingenuity had developed numerous machines for farming, but these bits of technology, the report explained, had only opened up more land on the plains to soil erosion, fertility depletion, and groundwater exhaustion. The committee thundered on, denouncing tenancy, decrying factory farms, and expressing mystification at how humans can believe in a future while they are gutting the land in the present. Water augmentation through reclamation projects was discounted as a chimera; drought was portrayed as a fact of plains life. The United States had to exploit the grasslands less if they were to continue to exploit them at all. The American love of the land was found wanting.

In the forty years since FDR had this report slapped on his desk several things have happened on the plains: growth has exploded, groundwater use has escalated, irrigated agriculture has flourished. Almost every stab by the committee at the future has proven incorrect. This exercise in planning was demolished by forces unleashed by the federal government and by forces the federal government seemed to know nothing about. While the national leadership fretted over the future of the grasslands, the humans in the sod dugouts on the High Plains were making a wonderful discovery. Deep well pumps could be powered by engines ripped from old Fords and Chevies.

Pump design had steadily improved through the twenties, and dependable models capable of high water volumes were available (Holcomb, W.H., 1929). The expense and woes of the oil-burning, belt-stretching powerplant were instantly overcome by hooking an old car motor on a direct drive train to the pump. Thousands of Americans in a generation of tinkering had made the internal combustion engine cheap, simple, and almost indestructible. The cost of installing an irrigation well (pump, powerplant, and drilling) dropped to $2,000 (Green, D.E., 1973).

Federal efforts to salvage the plains provided the capital for acquiring the new pumps and engines. The Agricultural Adjustment Act of 1933 flung money at the farmers. By the spring of 1934 one traveler on the High Plains stared in amazement at "the incongruousness of apparently new tractors and up-to-date farm equipment on farms where the farm people themselves are living in mere holes in the ground" (ibid.). During the first six months of 1935, farmers on the Llano Estacado bought a million dollars worth of tractors. Bankers got more capital from the federal Reconstruction Finance Corporation; as one explained, "There's not a damn way in the world I can lose any money" (ibid.). Government programs to support farm prices put a floor under this economic boom.

From 170 irrigation wells in 1930 the High Plains zoomed to 2,180 such wells (at least) in a decade. Irrigated fields went from a few thousand acres to a quarter million. As early as 1936 the president of Texas Tech University warned "that the water supply in our soil, while apparently abundant at the present time, can be depleted by injudicious use" (ibid.).

Response to the warning was swift. By 1948, the High Plains had 8,356 irrigation wells; nine years later this had grown to 42,225. During the fifties, two-thirds of the wells converted to the indigenous, cheap natural gas. Anhydrous ammonia produced by the region's petroleum supplies was sprayed on the land to compensate for declining fertility. By the sixties the High Plains had 5,500,000 acres under irrigation and men were working through the night to direct the flow from the ceaseless pumps. Fossil water and fossil fuels had made a billion dollar economy (ibid.).

There was one hitch. The Ogallala was running out of water. This came as a great surprise to many farmers exploiting the High Plains. For decades the humans of the Llano Estacado had believed the aquifer was inexhaustible for a simple reason. The Ogallala, they argued, was actually an underground river. Some thought it flowed from the front of the Rockies, some guessed it began up in the Arctic. The U.S. Geological Survey had explained as early as 1914 that the Ogallala was simply the storage of thousands of years of rain and snow (ibid.). Local people thought this conclusion preposterous.

The same year that federal experts were cutting the ground out from under the idea of inexhaustible water, Zenas E. Black, executive secretary of the Plainview Commercial Club, sketched the essential case for the endless Ogallala. "From whence does this water come?" he asked in 1914.

> There is only one way to answer it: The "Underground River," one of the largest systems in the United States, has been discovered. It starts in the melting snows of the Rockies, sinks below the surface and at the urge of gravity starts southeast . . .
>
> For hundreds of years this water has been flowing under the plains on its way to the Gulf and mankind knew it not. Now that this subterranean pipe-line is being tapped, the plainsmen claim that they have the nearest ideal system of agriculture on earth . . .
>
> There are no floods on the plains. And neither are there any drouths, for the "Underground Rain" always rises just when and in just the amount that the man who starts the pump desires. (Black, Z.E., 1914)

This belief endured well into the 1950s (Green, D.E., 1973). The farmers' stubborn persistence in asserting hydrological and geological nonsense is perhaps explained by the horrors of dryland farming. "I grew up on that dry-land stuff," remarked one successful plainsman. "I watched my daddy fight it and I fought it with him. A little cloud comes over and everybody sort of holds their breath and grunts, and it don't drop nothing and you just keep on watching the crops dry up. Uh-uh. When the water's gone I'm leaving too" (Graves, J., et al., 1971).

The Ogallala is sinking fast at current pump rates and will be commercially useless in a few decades. A lot of little things came together on the High Plains to produce one big thing, a billion dollar agricultural economy. This success, so different from the future predicted by the Great Plains committee in the thirties, was forged by cheap energy (cheap money, cheap water, cheap natural gas, cheap electricity, cheap fertilizer). Humans have escaped the buffalo world of the Nermeruh and the powerlessness of the early sodbusters by mining the fertility of the topsoil and the waters buried in the earth. This flight has led to a thing called the groundwater problem. Once again, people are putting forth a plan.

Plan

The original solution proposed for the imminent groundwater exhaustion of the High Plains was so straightforward it could be grasped by a child: build a pipe from the Mississippi River to the Llano Estacado and pump the water uphill. This project was part of a statewide Texas water plan which would siphon 12 to 13 million acre feet off the Mississippi each year and cost an estimated $8.966 billion to construct. Lifting the water 3,000 feet to the High Plains would have consumed 7 million kilowatts—equal to 40 percent of Texas electrical appetite in 1970 (Graves, J., et al., 1971; Jensen, C.W., and W.L. Track, 1973). Since the flat tableland offers no place for surface storage of the imported water, planners contemplated using the depleted aquifers. This plan ran into many problems (not the least of which was cost and scale) and is currently being revised.

Because the Texas constitution makes groundwater private property, some concern has been expressed about husbanding expensive, imported water if it is dumped into the Ogallala (Rayner, F.A., 1972). State water leaders also worry about getting the voters to

pay for the undertaking (High Plains Underground Water Conservation District, No. 1, Lubbock, Texas, 1970). Wresting the
water from the states bordering the Mississippi River is seen as no
simple task (Burleigh, H.P., 1970). Finally, should the water ever
get to the High Plains it promises to be so expensive (pumping
costs alone should run a fortune) that no one is sure the farmers
will be able to afford it (Graves, J., et al., 1971).[10] The inexhaustible Ogallala is going to be rescued by the inexhaustible Mississippi.

The initial plan responded to the growing belief on the High
Plains that the Ogallala would run out. Tom Milligan, editor of
the region's publication, *Irrigation Age*, made the point in the
1971 essay "Call the Law!" He recommended that farmers copy
the conservation practices of the oil industry "so a lush, green, and
rolling-in-money area like the High Plains area of Texas won't dry
up, blow away and return to a desert state." In Milligan's eyes the
choice was very simple: "If farmers, bankers, local businessmen,
manufacturers and others making their living off the land don't
take a giant step on land and quit irrigating all day and all night
even under the full moon, there just ain't gonna be a tomorrow for
a bunch of us" (Milligan, T., 1971). An economic study of the
region in the seventies said the same thing another way: "Groundwater and oil production are two of the basic resources for the
regional economy. The resources are exhaustible, and essentially,
non-renewable; depletion will require significant adjustments in
parts of the regional economy in the foreseeable future" (Osborn,
J.E., and W.C. McCray, 1972). To date the farmers have borrowed
one idea from the oil industry; in the 1950s they sued the Internal
Revenue Service and won a depletion allowance just like the oil-
men's (Graves, J., et al., 1971).

The Texas water plan, whatever its ultimate design, faces a bleak
future. Voters rejected a 1969 constitutional amendment that
would have set up a three billion dollar water development fund.
States now using the Mississippi don't think the river has any surplus water. Advocates of water importation worry that Americans
place "strong subjective value" on the Father of Waters and will
block any diminishment of its flow (Jensen, C.W., and W.L.
Track, 1973). In sum, the state contemplates seeking water it does
not own which it will move with money it does not have to a final
destination where it cannot control water use. This fate is in keep-

ing with modern hydrological thinking. Safe yield in west Texas has long been abandoned for economic growth.

Theoretically, mining an aquifer leads to a healthy economy, and a healthy economy can solve the problem of groundwater exhaustion. Aaron Wiener, in a recent book (1972), stated this position plainly:

> A case could possibly be made for keeping such stocks as a reserve for the needs of future generations or for some unspecified contingencies, and conceivably there might be situations where such an attitude would be justified. In the majority of cases, however, the opposite attitude, i.e., use of the resource by the present generation, will be the more meaningful one. Utilization of such one-time stocks will create new economic assets that might make it economically feasible either to replace in the future the then-exhausted supply by the import of water (natural or man-made) or, alternatively, to adopt a change in the resources base that would substitute other inputs for water or at least significantly reduce water requirements. . .

One might wonder on the basis of this argument whether there can ever be a groundwater problem. If resource exhaustion simply leads to yet another bounty, why worry? Clearly, in west Texas the consumption of Ogallala promises to be a springboard to economic ruin. But, it can be argued, this is an exceptional case.

Tucson, Arizona, is another case. Seventy-five years ago the city was not a city but a small town supplying local ranchers and miners. A little truck gardening persisted along the banks of the Santa Cruz River. "In the days of Old Tucson," explained George Wharton James in 1917, "the Santa Cruz Valley was regarded as unfit for farming land. When the rains came . . . it was fairly good for pasturage, but few of its earlier citizens ever regarded it as capable of being developed into productive farms" (James, G.W., 1917a). In 1912 investors from Chicago and London bought 12,000 acres in the valley, spent $30,000 surveying the aquifer, and then created Tucson Farms complete with ditches and electric pumps. "Crops never cease growing," James marveled. "There is no winter or dormant season" (ibid.). Fields that never went dormant never stopped drinking water.[11]

As a current British schoolbook explains, in places like Tucson "a kind of race begins" (Simons, M., 1967). Humans on the surface compete with each other in mining the aquifer. Prior to 1915, it is estimated that all the humans who had ever lived in the Santa Cruz Valley had withdrawn no more than 200,000 acre feet from the aquifer. From 1949 to the late sixties, pumping in the Tucson area grew enormously. At present this thirst has leveled at three to four times the rate of natural replenishment. The State Water Commission says the water table is dropping about 8 feet per year (Gerke, D., et al., 1975). The Tucson area is the state champion in mining groundwater.

Today, the city debates whether to participate in a one to three billion dollar water importation project from the Colorado River. Like west Texas, the cost of the water (when delivered) is a concern. Also, some people question if the once mighty Colorado holds enough water to permit such exports (U.S. Department of the Interior, 1974). The city of Tucson, which rests on ancient O-otam ground, abandoned safe yield, created a booming economy, and now has discovered that this economy may be a springboard to an exhausted aquifer and extremely expensive imported water (*if* water can be found to import).

Groundwater mining as a weapon against aridity raises more snakes than it kills. Planners have overlooked a couple of new facts. As cities and agricultural districts in the American West search for water to import they are increasingly finding that no one wants to export any. More humid regions often are using all their water for fisheries or simply to dilute and disperse industrial wastes. The Pacific Northwest and Canada assert that they need their heavy river flows as a reserve for their own future growth. The slow penetration of ecology into the discussion raises the question whether or not in an ecosystem any water is ever unused or wasted. A further difficulty lies in the key to moving water: energy. Cheap, convenient sources of concentrated energy are dwindling; thus, price estimates for moving water to central Arizona or west Texas must be continually revised upward. The arid and semi-arid regions that have spent the last few decades busily consuming their fossil groundwater now find that other regions have also been devouring resources, or have high hopes of doing so in the future. The dry-lands are reaching outward for resources that are simply not there.

The groundwater problem that nourishes so many academics, bureaucrats, and politicians is actually a groundwater reality. In arid lands groundwater is a nonrenewable resource, and hence, it is an example of nonrenewable resources in general. When such a resource is used by humans, it is gone, period. Hydrology, water management, and master plans cannot alter this fact.

For twenty or thirty thousand years humans have roamed and loved the American West. The long gone Hohokam canal builders, the silent dead of abandoned pueblos, the vanquished Nermeruh, the struggling O-otam, are part of this legion who attempted to make the region's resources pump energy and material faster and faster. Countless lives are buried in the efforts to make petroleum drive an engine, to find odd plants that would flourish beneath western skies. A host of lost geniuses tamed corn, learned to loot nitrogen from the air, doted on the pumps that now roar day and night. There has been no lack of dedication. The land harbors bones of padres like Kino, of Indians fighting to hold onto old ways, of Indian fighters killing to impose a new order. There has been no lack of courage. All this humanity has brought those now living to a situation difficult to misunderstand. Human numbers and human technology can use resources faster than geology, or rain, or sunshine can replace them. This is the ground and the water problem. We must now decide how to exploit the planet in the future. We must take care of business.

Part Three

Now I don't mind chopping wood,
And I don't care if the money's no good.
You take what you need and you leave the rest,
But they should never have taken the very best.

> *—J.R. Robertson,*
> *"The Night They Drove Old Dixie Down"*

It is the horses who make men wild. With a horse under me, I
myself am no better than a Comanche.

> *—New Mexican buffalo hunter, ca. 1860,*
> *explaining why he would not let his son ride*

1. Taking care of business

They say there are too many of us. They say the old live too long, the men and women have too many children. Not enough of the children die. And they go on and have more children, and all the men, the women, and the children want too much metal, eat too much food, combust too much energy, use too much water, spread over too much ground. Resource problems merely state the arithmetic of human numbers consuming fixed quantities beyond fixed rates of renewal.

Groundwater exploitation offers a stark example of this toting up of thirsts and buckets. The arid and semi-arid aquifers tend to be geologically discrete and meteorologically subject to low and slow recharge. Populations pumping such deposits are also clearly delineated. A piece of paper, a pencil, a few minutes, and numbers tumble forth telling how long the buried waters will last for how many people at such and such a rate. Of course the calculations are never perfect because aquifer size is always a rough guess. But the drift is unmistakable. Sinking water tables, like sinking bank accounts, deliver but one conclusion.

Moisture buried in the rock, sand, and gravel of the earth demonstrates the common nature of all resources in a manner that excites humans because humans can easily grasp the ingredients of the problem. In this instance, people can believe what is true of all materials: that they have a limited size, that they charge an energy cost for delivery, that different cultures impact different resources at different rates, that in a world where nothing is ever really gone almost anything can move beyond human reach. Groundwater is the perfect metaphor for studying the ways of different humans and the impact of different ways. The O-otam and Arizonan, the Comanche and the plainsman are easily divided by such phrases as stone age and industrial age, by such technologies as ollas and electric pumps. But thirst and the caloric demands of metabolisms bring them back together. In Pimería Alta and on the Llano Estacado enough has happened over the centuries to permit a look through the gauze of cultural myths into the machinery of human practices.

Concentration and energy cost

Humans often say they are baffled by notions of resource size and by statements about the energy costs of moving resources. Water circulates in a hydrological cycle and cannot leave the planet, so how can there be a water shortage? Energy can neither be created nor destroyed, so how can there be an energy crisis? Sometimes they ask a further question—they doubt the existence of a population problem. They point out that nations like Holland have a tremendous density of people and are prosperous while places like Arizona are barely inhabited. How can people use too much of something that is always circulating, combust too much of something that cannot be destroyed, or grow too numerous when some packed tracts of earth hold societies of plenty?

The planet cannot be water short, but humans and societies in portions of the planet can be. Chicago can be water short; New York can be water short; Tucson, Arizona, can be water short. Humid industrial cities can (and some do) use water for drink, for driving commercial processes, and for dispersing industrial wastes at a rate in excess of its delivery by rain, stream, well, and lake. When they do, beaches are closed on Lake Michigan; fish desert the Hudson. Humans are dependent upon water at certain concentrations and at certain purity. Since arid regions hold less water and that little water is physically outlined for easy measurements, they demonstrate this basic problem on a scale people can understand. The condition is not peculiar to arid lands, just more graphic in such places.

Humans are water short when they can not find enough moisture at a suitable concentration and purity for their appetites. Man demands rivers, not mists. The fact that the molecules of H_2O exist somewhere on the planet does not assuage thirst. The molecules must clump together in sufficient numbers and at appropriate places for humans to benefit. Aborigines required water holes and ephemeral arroyos. Early industrial Americans demanded rivers, lakes, and wells driven by wind. Modern technological societies must store surface flows behind gigantic dams and pump aquifers at high volumes to slake their thirst. Access to new concentrations of water in this century have been achieved by tinkering with the hydrological cycle. Engineering has built barriers to create new storages, and modern pumps have reached ancient storages in the earth previously little used.

Altering the hydrological cycle requires an investment of energy, and for nearly a century this energy has been found in fossil fuels. Coal, petroleum, and natural gas represent concentrations of sunlight. Like the thick topsoil early Americans found on the prairies and the deep aquifers modern Americans found in the arid basins, these resources result from storages of circulating things. On a planet of motion, some grasses trapped sunlight for a while at a rate in excess of animal predation and built up layers of rich dirt; in the same manner some soils soaked up rain and snow faster than the sun could evaporate it or plants pump it with roots. Plants and organisms perished in some places in quantities and at a rate exceeding any contemporary efforts to consume them; these organisms and plants were buried and concentrated into fossil fuels. They represent storages of sunlight largely lost to the life of the planet until recent times.

These storages are more easily consumed than replaced. The concentrations were achieved over long periods of time by geologic motion and solar energy. This is why economists have long considered them free goods: man did not have to pay for their concentration. When they are consumed, man will have to pay if he wishes to duplicate them. That is the root of the energy crisis, the groundwater problem, the various depletions of materials. It is not that energy, water, and minerals cease to exist when humans use them, but that they cease to exist in the previous concentrations.

The meaning of concentration can be made clear by looking at energy. Sunlight is not a simple substitute for coal, oil, and natural gas because it is more dilute than they are. A square meter of fossil fuels holds more calories, BTUs, and power than a square meter of sunlight. That is why solar collectors must cover a large area and why coal mines cover a small area. It is like the difference between a mist and a river. People build dams with turbines to convert the energy in rivers to useful power. Topography and solar energy collect this energy for them in drainages. Fossil fuels were gathered and collected in the same way over millions of years. When humans attempt to concentrate sunlight by their own contrivances they must pay this cost. They must round up the mist of sunlight into something approximating the lump of coal, cubic foot of gas, bucket of oil. This entails spending energy to concentrate yet more energy. So many calories must be invested in the solar collector to net so many calories in concentrated form. Things already collected (hydropower, fossil fuels) hold a competitive edge over things dilute (sunlight, wind power, tidal power, et al.).[1]

All proposed solutions to current resource problems lead one to energy, and all possible new sources of energy lead one to the facts of concentration and dilution. Humans, like all organisms, prefer particular concentrations of energy. Modifying coal, for example, to this level of concentration required mastery of potbellied stoves, boilers, and so forth. Mastery of nuclear power with massive concentrations of energy in a small mass has required decades of research and the erection of enormous powerplants shielding the population with thick walls. To exploit uranium (or plutonium in the yet unperfected breeder reactor) humans must invest enormous amounts of energy to control the power unleashed or they will die. That is what the talk of nuclear safety and nuclear expense means. In solar energy, wind power, and so forth, the problem is the reverse: the energy is so dilute that a huge energy investment is required to concentrate it. But both facets of the problem cut down the yield of such systems. Because of these facts, as fossil fuels dwindle in accessible concentrations power costs rise as man turns to other sources.[2]

Human cultures have been grab bags of tactics against these realities. Man has continually quested for concentrations of resources that he could afford. Comanches found a level they could master in the buffalo; later, Texans gained access to the Ogallala. One resource was renewable at the rates of use humans exercised, and one was not. That is what the problem is all about. The sun pouring through the grass and bison was concentrated at a rate no Indian seemed able to exceed. The waters in arid aquifers and in fossil fuel deposits are readily consumed by humans at rates far in excess of the speed at which rains and dying organisms replenish them. Any substitutes mean humans will take over the work of concentrating the water and concentrating the solar energy. Behind the water importation plans, the desalination plants, the cloud-seeding programs, the various alternative energy systems, is this reality.

Overpopulation simply means that humans in specific places cannot find the resources to sustain them at the moment. Holland is not considered overpopulated because it is able to import food, energy, and minerals for its people. India is cursed because it cannot. But, if one takes the stance that overpopulation means using resources faster than they are renewed within one's own national boundaries, then perhaps the United States is more overpopulated than Bangladesh. The poverty of Bangladesh is mainly the product of what its own land puts forth, and the prosperity of the United States is accomplished by plying 6 percent of the planet's humans with one-third of the planet's resources.

The debate

Some now call these facts of resource concentrations and the cost of concentrating resources the limits to growth. Humans gather in meetings and project current levels of consumption into a future of economic collapse. Many other humans refuse to believe these projections. They say something will turn up to remedy these hypothetical shortages. They say people cannot talk about limits to growth when much of the planet still lives in poverty. Or they say that these projections will prove a historical curiosity because humans will fling satellites into space and harvest the endless energy of the sun. But within the debate there is an actual consensus. No one really doubts the implications of the numbers. That is why responses to these numbers conjure up technological fixes while ignoring the facts of dilution and concentration, energy investment, and energy yield. That is why so many humans have been swiftly stricken with the idea of space colonies: by leaving the planet they hope to leave the arithmetic behind.

There probably is no escape from the facts of thermodynamics and the implications of using resources faster than they are replenished.[3] But even if there should be some rescue from these realities, it will not come in time to radically change the problems people will face in the next century. For decades to come growing numbers will be exploiting a declining resource base. In the case of groundwater in arid regions, this means its depletion will continue in an era when any methods of augmenting the declining supply will entail spending more money because of shrinking energy supplies. It means that its prime use to date, irrigated agriculture, will also have to endure rising costs. Modern irrigated agriculture uses fossil fuels in moving water, fertilizing ground, killing insects, destroying competing vegetation, and working the land with machines. In short, all the problems associated with exploiting fossil groundwater will grow worse because all the known solutions to these problems require concentrated energy, which will grow more costly.

There are various ways that humans can react to these facts of declining groundwater supplies. Numerous practices (trickle irrigation, lining ditches, etc.) can cut down water loss to the soil, air, and competing vegetation. Reinjection wells can pump drainage from agriculture back into the aquifer. Urban users can install toilets and showers more efficient in water use (or sanitation systems using no water) and abandon water-greedy landscaping for natural arid plants. Farmers can seek crops derived from arid vegetation that may prove more drought and insect resistant than the imported staples they now use. Given a constant population, a fixed cropland, a steady industrial base, and rigid, inflexible per capita use of water, these tactics will make a limited amount of water last longer. That is all. Unless use is reduced to recharge (the safe-yield condition), the aquifer will ultimately be dewatered. To date, populations in arid regions participating in industrial economies have shown little interest in such conservation measures. These methods cost the user more money (and energy) for the same amount of water than whole-scale exploitation of the aquifer. And they defer the benefits of exploitation to a later point in time. Humans have preferred the cheapest and swiftest way of mining groundwater.

Legal solutions to the problems of aquifer depletion have also aroused little enthusiasm. American and European law has tackled surface water for centuries, but ignorance of the hydrology of subsurface waters hindered their regulation until recent times. Now hydrological knowledge confronts fixed property interests. Since, in the arid American West, for example, humans have been held to own the water under their land, any new regulation of this right impinges on current assets. To legislate new limits on pump rates and insist that groundwater can only be used for certain ends dispossesses people of a resource that they previously owned without encumbrance. Some states have limited the number of new users who can pump from an aquifer. This tactic protects prior owners more than the aquifer. Government can confiscate groundwater as a group resource, but this will require the huge expense of indemnifying the previous owners. The costliness of such a solution, plus the fact that the economies now existent in the West are based on current levels of water depletion, has made the region as a whole reluctant to pass severe groundwater laws.

The marketplace seems an unlikely mechanism for husbanding groundwater. Economies (regardless of stripe) treat natural resources as free goods granted by nature. Success is measured by the rate at which these goods are extracted and dispersed. No economic register would suggest a gain if pumping on the High Plains of Texas was reduced to a safe-yield level. Nor would economic barometers consider it a windfall if the auto industry collapsed, though this would stretch the available supplies of metal and petroleum. Cheap groundwater is essential to growing arid economies because it enables them to compete with regions where water is more abundant and is delivered by rain.

To ask the economic sector to solve the problem of groundwater depletion is to ask it to self-destruct. There can be limited conservation and, in rare instances of high recharge and low use, the attainment of a safe-yield condition. But in the main, severely limiting the use of groundwater in arid lands means severely limiting the level of the economy. This change in economic activity can be good or bad—it depends upon what people want for themselves. People in arid regions must face up to the size of their water use and the reasons why the water table is sinking, and then decide what they plan to do about it. If they wish to consume the stored water swiftly, they can. If they want to stretch the supply out over the centuries, they can regulate use. But they would be wise to refrain from solving their current water-mining activities by pointing to some future solution which is currently unworkable and which promises to be unworkable in the future. Only by accepting that the well has a bottom can humans hope to use the contents judiciously.

Look again

The lay of the land tells us that resources in convenient concentrations are limited. In the case of some minerals, fuels, and water deposits, humans can see the point of exhaustion just decades ahead. Groundwater is the dominant factor determining the prosperity of modern economies in the arid lands. Access to groundwater triggers the use of other resources in such regions. Since recharge to aquifers in the drylands is slight, the more these water reserves are developed, the faster they will be depleted. When depleted, new supplies can be found through importation if humans are willing to pay the energy bill for moving the water, and if other humans are willing to permit exportation of their own water resources. Current trends make both possibilities increasingly unlikely.

The O-otam and kindred Piman people of the Sonoran desert afford an example of what living in balance with resources can mean, and what development can lead to. The humans trapped for centuries in a solar-fired world lived at peace with their water supplies, but they lived (at least to moderns) hard, impoverished lives. Safe from pollution, ignorant of aquifer drawdown, they experienced hunger, toil, motion, and group solidarity. Their access to resources helped determine what they thought resources were; they fashioned a unique view of the world. In their dances, beliefs, and efforts to gain new information, the possible implications of low-energy and low-water use are acted out. Facing peril, the O-otam demanded sharing; facing abundance, moderns leave the individual to himself.

In tracing the movement of the Piman people from the stick to the shovel to the tractor, a glimpse can be had of the price paid in natural resources for the feat described as development. The shift from rain to wells and from subsistence to a market economy changed the way humans lived and thought. Band and tribal solidarity dropped away until the isolated family was left. Piman lives lost contact with the resources of their own land and became defined by what loot individuals and families could gather from the American economy. Sharing surrendered to individual pursuits, and because of the workings of the national economy and the preparation of the Piman people for it, abundance was replaced with privation.

Groundwater development was crucial to the transition of the Piman people because water shortages were the glue that had held their societies together. The well undercut the reason for the group. Their experience is in many ways unique, but it is typical for the arid lands. The desert O-otam and riverine Pima have left balanced resource use for imbalanced use. They have in an impoverished way joined western society and thus become part of its groundwater problem: that urban, industrialized, and mechanized agricultural outposts of western man use groundwater in arid lands faster than it can be recharged by rain or snow.

This shift from renewable to nonrenewable resources has been accomplished in less time and to a greater degree on the High Plains of Texas. Where the O-otam stumbled for centuries toward the horse, the wheat, the pump, the shovel, the tractor, and the electric line, the distance between the last Nermeruh stand on the Llano Estacado and today is about a century. The Comanches offer another example of what peace with nature can entail, and their explosiveness upon acquiring the horse suggests the surges humans experience when sudden new flows of energy come within their reach. The resource anarchy of Comanche society was repeated among modern Americans when cheap power and improved pumps brought the Ogallala within reach following World War II. The spasms of High Plains development and the inability of humans to control and predict change on the grasslands temper modern hopes of tidily structuring new economies in third and fourth worlds. No one in the past foresaw the impact of the horse, the new pumps, and of federal aid on the grasslands.

The line connecting the O-otam, Comanches, Spaniards, sodbusters, and modern lords of the arid lands is the search for better pumps to aid in extracting resources: groups, wells, tools, fuels, mining techniques, machines. Past successes in this endeavor are the root of modern problems. We now talk of ecology, pollution, resource depletion, groundwater declines, and the carrying capacity of the land because ancestors found ways to exploit the land at great rates. We are becoming aware of what our history means.

The surprises of the past tell us to be careful in plotting out the future. Knowledge gained from the past warns us not to ignore the implications of declining reserves of concentrated resources, and not to beg the questions of the future by dreaming of some energy source free from the facts of thermodynamics. Growing human numbers and growing human appetites make any future solution to resource problems more enormous in quantity and more difficult in fact. To bring the rest of the human race up to American standards of living is a staggering task. To talk of American needs growing yet larger magnifies this challenge. Resources already spent and dispersed disadvantage the future; knowledge accrued from past exploitation may help the coming generations. We do not have to reinvent the wheel.

This writing has always been on the wall. It is not a revelation to learn that cheap energy makes societies boom, that groundwater in arid regions has negligible recharge, that humans tend to use as much of anything as they can lay hands on. We can ignore these facts and pump, mine, and combust with abandon, or we can recognize these facts and attempt to construct a sustainable society. There will be no painless answers, nor were there any in the past.

The O-otam and the Americans lived in two separate deserts.

Smithsonian Institution Photo No. 2788-D.

Smithsonian Institution Photo No. 2767-A.

O-otam home, summer village. Poles keep stock from eating
house. Smithsonian Institution Photos Nos. 2779-A, 2779-B,
2779-C.

House made of willow boughs. Smithsonian Institution Photo No. 2780.

Round grass house. Smithsonian Institution Photo No. 2779-H.

O-otam shelter: mesquite branches, grass. Smithsonian Institution
Photos Nos. 2784-C, 2784-D, 2784-E, 2784-F, 2784-G, 2784-H,
2784-I.

Smithsonian Institution Photo No. 2783-A.

O-otam granary of Pan Tak. Smithsonian Institution Photo No. 2781.

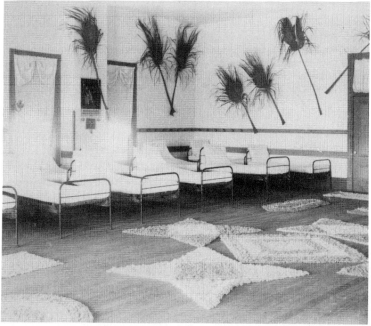

New place, new way: Phoenix Indian Industrial School, 1894.
Smithsonian Institution Photos Nos. 2786-D, 2786-F.

Notes

Part One

1. For example:

    > Water supplies were developed by inhabitants of ancient
    > cities many centuries before Christ. These supplies frequently
    > were secured from wells dug either in the solid rock or in
    > looser formation and walled with rock. These wells were from
    > 3 ft. to 9 ft. in diameter, and in some cases were more than
    > 200 ft. deep. Probably the most outstanding piece of ancient
    > engineering was Joseph's Well at Cairo, dug about the time
    > the pyramids were built, in the neighborhood of 3000 B.C.
    > This well was 18 ft. × 24 ft. square and 295 ft. in depth; two
    > lifts were used to pump the water, the upper being 160 ft.
    > and the lower 135 ft. The design of pump used was far re-
    > mote from those in service today and consisted of an endless
    > chain of buckets pulled by oxen. In order to get oxen down
    > to the 160 ft. level a spiral path was made around the upper
    > section of the well, a considerable achievement in itself.
    > (Holcomb, W.H., 1929)

2. Power over materials is a dream old, yet sharp, in the human mind.
    "Machines," one human noted, "may be made by which the
    largest ships, with only one man steering them, will be moved
    faster than if they were filled with rowers; wagons may be built
    which will move with incredible speed and without the aid of
    beasts; flying machines can be constructed . . . Machines will make
    it possible to go to the bottom of the seas and rivers . . ." (White,
    Jr., L., 1967). Such were the thoughts of Roger Bacon, A.D. 1260.

3. The Club of Rome report, *The Limits to Growth* (Meadows, D.H.,
    et al., 1972), is probably the best known study projecting current
    resource use into future resource exhaustion. But there are many
    similar works (Hubbert, M.K., 1969; Howe, C.W., et al., 1971;
    Lapp, R., 1973; Ford Foundation, 1974). For an economic study
    of the same bent see Georgescu-Roegen, N., 1971. James Ridge-
    way's *The Last Play* (1973) examines how one industry, energy,
    reacts to such projections. For a look at the environmental stress
    caused by development see *The Careless Technology* (Farvar, M.T.,
    and J.P. Milton, eds., 1972). These books do not so much predict

the future as demonstrate the inescapable consequences of continuing current rates of consumption. They sketch the fragile foundations of industrial societies.

4. For a sample of the literature on this point see Board on Agriculture and Renewable Resources, Committee on Agricultural Production Efficiency, J.G. Horsfall, chairman, 1975; Pimental, D., et al., 1973; Pimental, D., et al., 1975; University of California Food Task Force, 1974; Wittwer, S.H., 1975; Steinhart, J.S., and C.E. Steinhart, 1974; Odum, H.T., 1971; Rappaport, R.A., 1967; Georgescu-Roegen, N., 1975.

5. The notion of a nonrenewable resource seems to boggle modern minds. A few examples from a 1973 issue of *Daedalus* on growth:

> The economy is not a mindless glutton that will devour the last morsel before it notices that the plate is empty. Instead a variety of processes operate within it to allocate resources among alternative uses . . .
>
> The indisputable finiteness of the earth is, in fact, quite irrelevant to the questions of whether we must limit economic growth to prevent resource exhaustion. (Roberts, M.J.)

> When addressing this frequently heard concern, I suggest that one employ an expedient rule of thumb and simply ignore calculated dooms that are more than one hundred years away . . . If everything goes smoothly in the interim, the accelerated pace of scientific and technological progress will make our world and its problems almost unrecognizable a hundred years from now. It is hard to conceive of a growth-induced disaster that could not be averted by some significant scientific innovation. (Zeckhauser, R.)

> The key argument against a larger population size hinges on the concept of a nonrenewable resource which would either be used up at a more rapid rate or have to be shared among more people. It is difficult to come to grips with this proposition because of the slipperiness of the notion of nonrenewability. For example, clean air and water cannot be regarded as nonrenewable, despite the current decline in availability. There is obviously a price tag on these goods which heretofore we have more or less treated as free; our ability to pay that price, however, depends on our productivity . . . (Ryder, N.B.)

6. When Europeans came within reach, Seris raided them. The Seris were practically exterminated in the late nineteenth and early twentieth century. Today, the few hundred survivors scrape by with a little fishing and by peddling ironwood carvings. Their

ancient lands west of Hermosillo, Sonora, have been made a vast irrigation district by the Mexican government. Based on ground-water, the area is booming and contending with a declining water table and salt water intrusions into the aquifer.

7. The omnivorous Seri held rodents a taboo and would not eat or kill them. Their lands were almost devoid of reptiles but overrun with rats and mice. As a consequence the terrain was honeycombed with burrows, which Mexican horsemen in hot pursuit of Seri raiders found all but impossible to traverse. Inadvertently, rodents wound up in the first line of Seri defense (McGee, W.J., 1971).

8. Primitive food-gathering systems are being reexamined by scholars. It has been found that some surviving hunting and gathering groups are characterized by stable populations, sound diet, and abundant leisure. One group, the Kung!, roam the Kalahari desert, consume 2,100 calories per day, and spend only three days a week gathering food. Similarly, it has been discovered that a family in the Middle East could gather enough wild wheat in three weeks to last a year (Ucko, P.J., and G.W. Dimbleby, eds., 1969). For similar observations see Boserup, E., 1965; Rappaport, R.A., 1967; Steinhart, J.S., and C.E. Steinhart, 1974. One demographer doubts that explosive population growth could have shattered such economies in the past (Petersen, W., 1975). The various studies turn an old outlook on its head: maybe man was not expelled from the garden but rather driven into it. For speculation on the origin of agriculture in the Sonoran desert see McGee, W.J., 1895.

9. For a vivid description of the back-breaking labor required by pre-European desert agriculture see S.P. Clark (1928) on Hopi farming.

10. Kino thought big thoughts. Consider his 1703 letter to the viceroy:

> Ten years ago I finished the building of a large church and residence in this town of Nuestra Señora de los Dolores. The churches of the other towns . . . are in the course of construc-tion . . . Father Agustin de Campos is also building three churches and three residences in as many centers entrusted to him. The six towns mentioned have sufficient property and means to support them: herds of cattle, and sheep, droves of horses, and fields with their abundant harvests of wheat, corn, beans and produce.
>
> Here at Nuestra Señora de los Dolores we have a well-built and spacious church artistically furnished and provided with good linens and altars, and seven bells. The mission also boasts beasts of burden, a water-powered mill, carpenter and blacksmith shops, oxen for plowing, cultivated land, produc-

tive gardens, a vineyard sufficient for mass wine, trees that
supply us with abundant Castilian fruit, and so on . . . I have
been able to undertake during these fifteen years some forty
expeditions . . . into unexplored regions . . .

. . . The land is fertile and of as high a quality as the best
in Europe. It would suffice . . . to order that twenty soldiers
from the nearby garrisons in Sonora merely make the rounds
once or twice a year of the twelve mission centers . . . And
then as soon as the Pimería province has been properly
organized, thought should be given—as can readily be done—
to the conversion of the other tribes . . .

On the banks of the bounteous and fertile Rio Colorado,
close to the head of the Sea of California, at the 32nd degree
of latitude, a town of some 300 or 400 families should be
founded and maintained. This could be done with a moderate
initial outlay that would render unnecessary any exorbitant
expenses resulting from a yearly subsidy . . . We missionaries
and the neighboring colonists of the nearby province of
Sonora would be able to assist these families with herds of
cattle and sheep and droves of horses. To support them, there
are very fertile fields of wheat and corn, and as vast a variety
of vegetables as in Europe, and other produce.

And in time an entrance could be made into the nearby
country of the Moquis (Hopi), and along the northern coast
to the regions known as the Gran Quivira and the Gran
Teguayo, as far as Cape Mendocino and the land called Yeso;
and following the north western and western coastline even
as far as the territory close to Japan; or going northeast and
east, one can reach the regions above New Mexico, and thus
establish a line of communication and trade with New France;
and if continued on to Europe, this latter route would be less
than half that via Vera Cruz . . . (Burrus, E. J., ed., 1961)

11. Witness this report filed by Fray Antonio Ramos on the missions
of the Santa Cruz valley in 1774:

Concerning the fifth point, be noted that the Indians in these
pueblos very often absent themselves wandering through the
various missions in search of food. There are two causes for
this: the first, and principal one, is their laziness and indo-
lence, which is common to every Indian. And the second
cause is that the lands of these pueblos yield very little. Even
though at planting time they get crops into the ground then,
as they are accustomed to wander, they leave them unattended.
Finally, they return, each getting but little from a limited
harvest. This in the minds of prudent people brings about a

consideration of many spiritual and temporal evils that follow from said wandering . . . (Baldonado, L., 1959)

12. For the Apache point of view of this commerce see Basso, K.H., ed., 1971.

13. The best ethnographic work on the western Apache bands is Goodwin, G., 1969.

14. Powell's 1878 *Report on the Lands of the Arid Region of the United States* systematically reviews and destroys massive development schemes for the arid west. Henry Nash Smith's *Virgin Land* (1950) explores the ideas of nineteenth century Americans on the exploitation of the west.

15. There is no agreement among scholars on bovine impact on the west. See James R. Hastings and Raymond M. Turner, *The Changing Mile* (1965) for a thorough review of the argument concerning the Sonoran desert.

16. H.V. Clotts' federal survey in 1915 exhibits the belief of officials that at last they had found an unbroken group of native Americans:

> They are well built of moderately large stature, strong of feature and personality, with a quiet, dignified bearing, which indicates ability to think and act for themselves and to meet with ingenuity all problems which may present themselves concerning the preservation of life and pursuit of happiness on the desert . . .
>
> It appears that the greatest need of the Indians is for wells and deeper, cleaner charcos. If wells were located at various points throughout the valleys, the necessity for moving to the mountains each dry spell would be eliminated. Their cattle could be grazed over more territory and the total number thereby increased. In addition, gardens or small tracts of land could be irrigated . . . (Clotts, H.V., 1917)

17. The resentment was plainly expressed in O-otam fashion. "Even today," one observer wrote in the late 1940s, "when the Papago expect a visitor from the Agency whom they fear or dislike, it not rarely happens that they leave their houses and hide in the mountains until they feel they are safe again . . ." (Bauer, R.W., 1971).

18. Estimates of energy reserves must be taken with a grain of salt. Exponential growth in consumption and rising costs in extraction easily shrink them. "And however abundant the earth's reserves of fossil fuels and minerals," E.J. Mishan has pointed out, "their continued mining must eventually exhaust them, and faster than we are prone to imagine . . . If, for example, actual oil reservoirs turn out to be four times as great as we estimate today, the current growth of consumption could be maintained for only fifteen years longer than it could according to current estimates" (1973).

Part Two

1. For the raw numbers of dollars and water on the High Plains see Jensen, C.W., and W.L. Track, 1973; Peckham, R.C., et al., 1967; Vandertulip, J.J., L.B. Seward, and J.T. Thornhill, 1966; Osborn, J.E., and W.C. McCray, 1972; Hansen, H., 1969; Alexander, Jr., W.H., 1961; Breeding, S.D., et al., 1964; Alexander, Jr., W.H., and J.W. Lang, 1945. For a look at exploitation problems on the High Plains exclusive of Texas see MacPhail, D.D., ed., 1972. Much of the following discussion of man's efforts to make a modern economy on the Texas High Plains is derived from one book, Donald Green's *Land of the Underground Rain* (1973). It examines the resource floor underneath the history of the American west (in this instance an aquifer), and in this it is almost unique.

2. The standard volume for Comanche ethnography is Wallace, E., and E.A. Hoebel, 1952. A recent popular history of the tribe, Fehrenbach, T.R., 1974, is the most substantial one to date. Comanche habits during their heyday on the Llano Estacado limited Spanish, Mexican, and American observations. There is a fascinating literature of captive narratives penned by survivors of Comanche kidnapping.

3. Plains Indian use of bison was as follows: HIDE—moccasin tops, cradles, winter robes, bedding, breechclouts, shirts, leggings, belts, dresses, pipe bags, pouches, paint bags, quivers, tipi covers, gun cases, lance covers, coup flag covers, dolls; RAWHIDE—containers (for clothing, headdresses, food, medicine bags), shields, buckets, moccasin soles, rattles, drums, drumsticks, splints, cinches, ropes, thongs, saddles, stirrups, knife cases, bull boats, quirts, armbands, lance cases, horse masks, horse forehead ornaments, bullet pouches, belts; HAIR—headdresses, saddle pad fillers, ropes, pillows, ornaments, halters, medicine balls; HORNS—cups, fire carriers, powderhorns, spoons, ladles, headdresses, signals, toys; TAIL—medicine switches, fly brushes, lodge exterior decorations, whips; SKIN OF HIND LEG—moccasins or boots; HOOF AND FOOT—glue, rattles; MEAT—everything eaten with guts devoured on the spot; SKULL—ceremonies, prayer; BRAINS—used in hide preparation; TONGUE—favored cut; BEARD—ornaments for dresses or weapons; SCROTUM—rattles; MUSCLES—glue, sinews for thread, bowstrings; FECES—fuel, signals, ceremonial smoke; BONES—knives, arrowheads (ribs), shovels, splints, winter sleds, arrow straighteners, saddle trees, war clubs, scrapers (ribs), quirts, awls, paint brushes (hipbones), game dice; BLADDER—sinew pouches, quill pouches, small medicine bags; PAUNCH—lining used for buckets, cups, basins, dishes; FOUR-CHAMBERED STOMACH—first stomach, medicine for frostbite and skin diseases; liner, a container for carrying

and storing water or a cooking vessel; WHOLE ANIMAL—totem, clan symbol (Mails, T.E., 1972).

4. Available concentrated energy defines the potential of a society but does not dictate what the society will do with this potential. For example:

> By 1850 anyone could see that the revolution in productive technology enabled fewer people to manage more resources. Instead of utilizing this fact and limiting population growth, thereby providing themselves with a utopian level of living with little effort, the Europeans used it overwhelmingly to support more people. Europe became the world's most densely settled continent and, in turn, sent out migrants to overrun whole new continents . . . [The] human species is now . . . using an extremely advanced technology to maintain four billion people at a low average level of living while stripping the world of its resources, contaminating the water, soil, and air, and driving most other species into extinction, parasitism, or domestication. (Davis, K., 1973)

5. Recent scholarship suggests that buffalo are more suited to plains vegetation than "speckled cattle." D.G. Peden, et al., in "Trophic Ecology of *Bison Bison L.* on Shortgrass Plains" (1974) concluded: "Bison have a greater preference for warm-season grasses and appear to feed less selectively than cattle . . . Sheep consume fewer grasses than either species. Bison appear to have a greater digestive power than cattle when consuming low protein, poor quality forage . . . These two mechanisms of feeding strategies may permit bison to exploit more fully than cattle the herbage resources on shortgrass plains."

6. Generations of Americans since the destruction of the plains tribes have found it difficult to let go of their memory. Films dwell on the horse people. Wounded Knee, an episode in their long death, still tugs at American consciousness. Poets wander in the fabric of dead buffalo economies. For example, two poems from Joe Ribar's volume *The Book of the Buffalo* (no date, no publisher):

> I've seen it all it all
> where once great herds grazed
> at sunset
> someone built a public
> school
> and children
> swing on swings
> dreaming at night of sugarplums

> in houses built
> on buffalo dust
>
> just beautiful
>   beautiful
>   buffalo
> no V-8's
> no america

7. Windmills are having a certain vogue at present (for example, Reynolds, J., 1970). The experience of earlier Americans with this "free" source of energy offers a handy note of reality to this modern interest: Wolff, A.R., 1890; Murphy, E.C., 1897; Hutson, W.P., 1898; Sageser, A.B., 1967; Hood, O.P., 1898. Memoirs of pioneers can also be culled for the delights of repairing such devices.

8. Sensing the impact such a pump could have on the Ogallala aquifer, Black railed against "disconcerting scientists [who] rave about running out of this or that article . . ." (Black, Z.E., 1914).

9. Shortly after the Puritans stepped off the boat in Massachusetts Bay, the frontier became an obsession and problem for Americans. John Winthrop, the first governor of that early colony, despaired because "many crept out at a broken wall." What he meant was that the group's dream of a city upon a hill was mauled and broken by the land, the continent of virgin land. Jefferson's small manageable republic of yeomen was bloated by the land. All the efforts at setting aside Indian nations were broken by the land. Westering became a fact of national experience, and a panacea for national woes. The homestead would be a safety valve for festering cities. Failures in the East were advised to go West, go West. When Aaron Burr cut down Alexander Hamilton and shattered his career and reputation, he went West.

When the census reports, surveys, and federal statistics in the late nineteenth century piled up, young scholars like Frederick Jackson Turner put forth a frontier hypothesis of American history. The idea was simple: the American people were the product of going West into the trees, prairies, plains, and mountains, and with these gone, what would the American people become? In the scholarly world, they became, in good part, buffs searching the entrails of their conquest of the West for anecdote and solace. Much of what is written as western history is a tale, told and retold— a sort of creation myth treasured by the American people.

Two men have tackled this question of what going West meant in different but mutually compatible ways. William Appleman Williams has written a string of histories on American foreign policy that find the roots of American economic and physical expansionism in the lessons learned on the long frontier. To be contained was thought to be slowly starved and defeated. Thus, within a decade of Turner's stunning presentation of his frontier hypothesis, an American Secretary of State demanded an Open Door in China. By the late 1940s American statesmen were convinced they could break the Soviet Union if they could just freeze its frontiers: containment. Americans associate growth with prosperity, survival, and the very well-being of the nation.

Walter Prescott Webb focused on what the resources of the American West (and the Western hemisphere) meant to human beings. In the *Great Plains* and the *Great Frontier* he found they meant riches, and in the United States a form of mass democracy unique on the planet. He wrote:

> . . . scholars will look back on the age when the Golden Door opened, and men marched out to the Great Frontier to create the greatest boom that the world has known; they will make myths and legends about it, and in poetry express their poignant yearning for New Frontiers. They will see the frontier as the great factor in the age called modern, see it clearly as the lost factor which they would so love to find. *Great Frontier*, pp. 27–28)

> If you could hold in your right hand the earth in miniature as it was in 1500 or 1600 and in your left hand the earth as it is now, which earth would you consider richer in resources? Or preferable as a desirable base of future operations? (Ibid., p. 292)

These two men have raised the questions that Americans have to answer to understand their past, present, and future. Though hardly mentioned in this essay, their works are the ground which has nourished it.

10. The importation scheme has been studied extensively: Vandertulip, J.J., L.B. Seward, and J.T. Thornhill, 1966; Jensen, C.W., and W.L. Track, 1973; Peckham, R.C., et al., 1967. For a dissenting view, see Graves, J., et al., 1971.

11. Now that western rivers are largely dammed, and the aquifers exploited, a scholarship is appearing that wonders if the game was worth the candle. The gist of the debate can be found in Berkman, R.L., and W.K. Viscusi, 1973.

Part Three

1. For a full discussion of this business of concentration and dilution by men in two different disciplines see Odum, H.T., 1971, and

Georgescu-Roegen, N., 1975. The matter is usually called thermo-dynamics. For a wonderful and tantalizing speculation about its implications see Henry Adams, *A Letter to Teachers of American History* (1910).

2. The same factors bite into the net yield of fossil fuels as early, easy-to-get deposits disappear. Coal at depth costs more calories to mine than coal close to the surface. Oil locked in oil shale costs more calories to extract than oil found in liquid pools. In some instances, the cost may be intolerable. This will mean such reserves will remain inaccessible to humans.

3. For an example of one such proposed escape from limited re-sources, space colonization, see O'Neill, G.K., 1974a and 1974b; Ehricke, K.A., 1971. Technological fixes restricted to the planet's surface are fusion power and breeder reactors: by offering abun-dant power (the hope of their advocates) they offer the possibility of concentrating and moving resources at low cost. For discussions of these systems see Bowden, C., 1975. Solar energy systems based on the earth's surface also have strong supporters (again, see Bowden, C., 1975). All these proposals lose some luster when examined in the light of net energy yield, the cost of containing by-products, and the timetable in which they can realistically be implemented. And, of course, the growing population forces such proposals to satisfy expanding appetites.

Bibliography

Adams, A.D.
    1910   Electric pumping for irrigation. *Electrical World* 55(26):
           1709–1711.
Adams, H.
    1910   *A letter to teachers of American history.* Press of J.H. Furst,
           Co., Baltimore.
Adams, R.N.
    1975   *Energy and structure: a theory of social power.* University of
           Texas Press, Austin.
Agency for International Development
    1975   AID in an interdependent world. *War on Hunger: a report
           from the Agency for International Development* 9(6):
           1–18.
Alexander, Jr., W.H.
    1961   *Geology and groundwater resources of the northern High
           Plains of Texas.* Progress Report No. 1, Bulletin 6109, Texas
           Board of Water Engineers.
Alexander, Jr., W.H., and J.W. Lang
    1945   *Groundwater in High Plains of Texas.* Progress Report No. 5,
           Texas Board of Water Engineers.
Alonso, W.
    1973   Urban zero population growth. *Daedalus* 102(4): 191–206.
Ambroggi, R.P.
    1966   Water under the Sahara. *Scientific American* 214(5):
           291–299.
Bahr, D.M.
    1964   *Piman social organization and resource management.* Arizona
           State Museum, Tucson. [Typescript.]
Bahr, D.M., et al.
    1974   *Piman shamanism and staying sickness: Ká:cim múmkidag.*
           University of Arizona Press, Tucson.
Bainbridge, J.
    1972   *The super-Americans: a picture of life in the United States
           as brought into focus, bigger than life, in the land of the
           millionaires—Texas.* Holt, Rinehart, & Winston, New York.
Baldonado, Luis.
    1959   Missions San Jose de Tumacacori and San Xavier del Bac in
           1774. *Kiva,* 24(4): 21–24.

Barbour, Erwin H.
  1899 *Wells and windmills in Nebraska.* Water Supply and Irrigation Papers of the U.S. Geological Survey, no. 29. 55th Congress, 3rd Session, Serial No. 3815, House Document No. 299, Washington, D.C.
Basso, K.H., ed.
  1971 *Western Apache raiding and warfare.* University of Arizona Press, Tucson.
Bauer, R.W.
  1971 The Papago cattle economy: implications for economic and community development in arid lands. In *Food, fiber and the arid lands,* ed. W.G. McGinnies et al., pp. 79–102. University of Arizona Press, Tucson.
Beals, R.L.
  1943 *The aboriginal culture of the Cahita Indians.* Ibero-Americana: 19. University of California Press, Berkeley.
  1945 *The contemporary culture of the Cahita Indians.* Bulletin 142, Bureau of American Ethnology, Washington, D.C.
Berkman, R.L., and W.K. Viscusi
  1973 *Damning the West.* Grossman Publishers, New York.
Black, Z.E.
  1912 The pump in the South Plains. *Earth* 9(March): 13–14.
  1914 The land of the underground rain. *Earth* 11(April): 13–14.
Board on Agriculture and Renewable Resources: Committee on Agricultural Production Efficiency, J.G. Horsfall, Chairman.
  1975 *Agricultural production efficiency.* National Academy of Sciences, National Research Council, Washington, D.C.
Bocking, R.C.
  1972 *Canada's water: for sale?* James Lewis & Samuel, Publishers, Toronto.
Bolton, H.E., ed.
  1919 *Kino's historical memoir of Primería Alta.* Vol. 1. Arthur Clark Co., Cleveland.
  1936 *Rim of christendom: a biography of Eusebio Francisco Kino, Pacific coast pioneer.* MacMillan Co., New York.
Boner, H.A.
  1971 *Hungry generations: the nineteenth century case against Malthusianism.* Russell & Russell, New York.
Boserup, E.
  1965 *The conditions of agricultural growth: the economics of agrarian change under population pressures.* Aldine, Chicago.
Boulding, K.E.
  1973 The shadow of the stationary state. *Daedalus* 102(4): 89–102.

Bowden, C.

1975 *Impact of energy development on water resources in arid lands.* Office of Arid Lands studies, University of Arizona, Tucson.

Breeding, S.D., et al.

1964 *Fifty years of water development in Texas.* Bulletin 6403, Texas Water Commission.

Briggs, P.

1972 Coping with population growth and a limited resource. Proceedings of the 40th Annual Meeting of the Western Snow Conference, April 18–20, 1972, Phoenix, Arizona. Printed by Colorado State University, June–August 1972, Fort Collins.

Brockett, L.P.

1882 *Our western empire: or the new west beyond the Mississippi.* Bradley, Garretson, & Co., Philadelphia.

Brooks, H.

1973 The technology of zero growth. *Daedalus* 102 (4): 139–152.

Brown, Jr., G.W., ed.

1968 *Desert Biology,* Vol. 1. Academic Press, New York.

1974 *Desert Biology.* Vol. 2. Academic Press, New York.

Brown, J.W.

1975 Native American contributions to science, engineering, and medicine. *Science* 189(4196): 38–40.

Brown, L.

1973 Rich countries and poor in a finite, interdependent world. *Daedalus* 102(4): 153–164.

Brunne, G.

1970 The Texas Water Development Board cooperative studies of the Ogallala underground reservoir. In *The Ogallala Aquifer—a symposium*, pp. 227–242. Special Report No. 39, International Center for Arid and Semi-Arid Land Studies, Texas Tech University, Lubbock.

Bryan, K.

1922 *Erosion and sedimentation in the Papago country, Arizona.* U.S. Geological Survey Bulletin No. 730, Washington, D.C.

1925 *The Papago country.* U.S. Geological Survey Water Supply Paper No. 499, Washington, D.C.

1941 Pre-Columbian agriculture in the southwest, as conditioned by periods of alluviation. *Annals, Association of American Geographers* 31:219–242.

Burdon, D.J.

1971 Exploitation of groundwater for agricultural production in arid zones. In *Food, fiber, and the arid lands,* ed. W.G.

McGinnies et al., pp. 289–300. University of Arizona Press, Tucson.

Bureau of Indian Affairs, Papago Agency.
1972 *Facts about the Papago Indian reservation and the Papago people.* U.S. Public Health Service.

Burleigh, H.P.
1970 Export of Mississippi River water to Texas and New Mexico. *Journal of the American Water Works Association* 62(6): 367–375.

Burnham, D.
1975 Bright hopes for N-power supply dim: reactor costs sky-rocket. New York Times News Service, *Arizona Daily Star,* Tucson, November 17, sect. A, p. 1.

Burrus, E.J., ed.
1961 *Kino's plan for the development of Pimería Alta, Arizona and Upper California.* Arizona Pioneer Historical Society, Tucson.

Cantor, L.M.
1967 *A world geography of irrigation.* Oliver & Boyd, Edinburgh, Scotland.

Carlson, M.E.
1969 William E. Smythe: irrigation crusader. *Journal of the West* 7:41–47.

Carr, Jr., J.T.
1966 *Texas droughts: causes, classification, and prediction.* Report No. 30, Texas Water Development Board, Austin.

Carter, G.F.
1945 *Plant geography and culture history in the American Southwest.* Viking Fund Publications in Anthropology, No. 5. Wenner-Gren Foundation for Anthropological Research, New York.

Casey, H.E.
1972 *Salinity problems in arid lands irrigation: a literature review and selected bibliography.* Office of Arid Lands Studies, University of Arizona, Tucson.

Castetter, E.F., and W.H. Bell
1942 *Pima and Papago Indian agriculture.* University of New Mexico Press, Albuquerque.
1951 *Yuman Indian agriculture: primitive subsistence on the lower Colorado and Gila rivers.* University of New Mexico Press, Albuquerque.

Childs, T.
1954 A sketch of the "Sand Indians." *Kiva* 19(2–4): 27–39.

Clark, S.P.
1928 Lessons from southwestern Indian agriculture. *Bulletin,*

University of Arizona Agricultural Experiment Station,
no. 125, pp. 233–252.

Clotts, H.V.

1915   *Report on nomadic Papago surveys.* U.S. Indian Service,
Department of Interior. [Typescript in Arizona State Museum
library, Tucson.]

1917   *History of the Papago Indians and history of irrigation of
Papago Indian reservations, Arizona.* U.S. Indian Service,
Department of Interior. [Typescript in Arizona State Museum
library, Tucson.]

Cloudsley-Thompson, J.L.

1965   *Desert Life.* Pergamon Press, Oxford.

Conkling, H.

1934   The depletion of underground water supplies. *Transactions of
the American Geophysical Union* 15:531–539.

Conselmen, F.H.

1970   The significance of the Ogallala formation in Texas. In
The Ogallala Aquifer—a symposium, pp. 2–4. Special Report
No. 39. International Center for Arid and Semi-Arid Land
Studies, Texas Tech University, Lubbock.

Cottrell, F.

1969   Organic energy and the low-energy society. In *Sociological
theory,* ed. W.L. Wallace, pp. 145–160. Aldine, Chicago.

Crosson, P.R.

1975   Institutional obstacles to expansion of world food production.
Science 188(4188): 519–523.

Daedalus

1973   The no-growth society. *Daedalus* 102(4): 1–253.

Davidson, E.S.

1973   *Geohydrology and water resources of the Tucson basin.*
U.S. Geological Survey Water Supply Paper No. 1939-E,
Washington, D.C.

Davis, A.P.

1908   Reclamation of the arid west by the federal government.
American Academy of Political and Social Science
31:203–218.

Davis, K.

1973   Zero population growth: the goal and the means. *Daedalus*
102(4): 15–30.

Davis, S.N.

1974   Hydrogeology of arid regions. In *Desert biology,* vol. 2,
ed. G.W. Brown, Jr., pp. 1–31. Academic Press, New York.

Densmore, F.

1929   *Papago music.* Bureau of American Ethnology, Washington,
D.C.

Dobie, J.F.
　1935　*Tongues of the Monte*. Little, Brown & Co., Boston.
Dobyns, H.F.
　1949　*Report on investigations on the Papago reservations*. Cornell University, Department of Sociology-Anthropology, Ithaca, New York.
Doelle, W.
　1975　*The adoption of wheat by the Gila Pima: a study in agricultural change*. Department of Anthropology, University of Arizona, Tucson.
Dooge, J.C.I.
　1973　The nature and components of the hydrological cycle. In *Man's influence on the hydrological cycle*. Irrigation and Drainage Paper, Special Issue No. 17, pp. 1–18. Food and Agricultural Organization of the United Nations, Rome.
Downing, T.E., and M. Gibson, eds.
　1974　*Irrigation's impact on society*. Anthropological Papers of the University of Arizona Press, No. 25. University of Arizona Press, Tucson.
Dregne, H.E.
　1969　*Prediction of crop yields from quantity to salinity of irrigation water*. New Mexico Agricultural Experiment Station Bulletin 543.
Dresher, W.H.
　1975　Technology—the future of raw materials supply. *Field Notes from the Arizona Bureau of Mines* 5(2): 1–3, 9–10.
Drower, M.S.
　1954　Water supply, civilization and agriculture. In *A history of technology*, vol. 1, ed. C. Singer et al. Oxford University Press, New York.
Dunbier, R.
　1968　*The Sonoran desert: its geography, economy and people*. University of Arizona Press, Tucson.
Dyson-Hudson, N.
　1972　The study of nomads. *Journal of Asian and African Studies* (Leiden, Netherlands) 7(1–2): 2–29.
Dyson-Hudson, R.
　1972　Pastoralism: self-image and behavioral reality. *Journal of Asian and African Studies* (Leiden, Netherlands) 7(1–2): 30–47.
Eaves, C.D., and C.A. Hutchinson
　1952　*Post City, Texas: C.W. Post's colonizing activities in West Texas*. Texas State Historical Association, Austin.

Ehricke, K.A.
    1971   The extraterrestrial imperative. *Bulletin of Atomic Scientists*
         27(November): 18–26.
Electrical World
    1910   Electricity in irrigation. *Electrical World* 56(3): 150–152.
Ellis, F.H.
    1970   Irrigation and water works on the Rio Grande. Pecos Confer-
         ence Water Control Symposium, May, 1970. Arizona State
         Museum, Tucson. [Xerox.]
El-Zur, A.
    1965   Soil, water and man in the desert habitat of the Hohokam
         culture. Ph.D. dissertation, University of Arizona, Tucson.
Engelbert, E.A., ed.
    1965   *Strategies for western regional water development: proceed-
         ings of the western interstate water conference, Corvallis,
         Oregon, 1965.* Water Resources Center, Los Angeles.
Ennis, Jr., W.B., W.M. Dowler, and W. Klassen
    1975   Crop protection to increase food supplies. *Science* 188(4188):
         593–597.
Evenari, M., L. Shanan, and N. Tadmor
    1971   *The Negev: the challenge of a desert.* Harvard University
         Press, Cambridge.
FAO/UNESCO
    1973   *Irrigation, drainage, and salinity: an international sourcebook.*
         Hutchinson & Co., London.
Farvar, M.T., and J.P. Milton, eds.
    1972   *The careless technology: conference on the ecological aspects
         of international development.* Natural History Press, Garden
         City, N.Y.
Fehrenbach, T.R.
    1974   *Comanches: the destruction of a people.* Alfred A. Knopf,
         New York.
Felger, R.S.
    1975   Nutritionally significant new crops for arid lands: a model
         from the Sonoran desert. In *Priorities in child nutrition in
         developing countries,* ed. J. Mayer and J.W. Dyer, pp. 373–
         403. United Nations' International Children's Fund, New
         York.
Flannery, K.V.
    1965   The ecology of early food production in Mesopotamia. *Science*
         147(3663): 1247–1256.
Flannery, K.V., and M.D. Coe
    1964   Microenvironments and Meso-American prehistory. *Science*
         143(3607): 650–654.

Fontana, B.L.
 1957   *Notes for Pima Indian claims case.* Arizona State Museum,
         Tucson. [Xerox of typescript.]
 1958   A detailed history of the Pima Indians of Arizona, with special
         emphasis on their location and history of their water supply
         —between 1846 and 1883. Report prepared for R.A. Hacken-
         berg for use in Pima Indian Claims Case. Arizona State
         Museum, Tucson. [Xerox of typescript.]
 1960   Assimilative change: a Papago Indian case study. Ph.D. dis-
         sertation, University of Arizona, Tucson.
 1963   Pioneers in ideas: three early southwestern ethnologists. *Jour-
         nal of the Arizona Academy of Science* 2(3): 124–129.
 1964   *The Papago Indians.* Arizona State Museum, Tucson. [Xerox
         of typescript.]
 1974   Man in arid lands: the Piman Indians of the Sonoran desert.
         In *Desert biology,* vol. 2, ed. G.W. Brown, Jr., pp. 489–529.
         Academic Press, New York.
 1975   The desert domain: people and land in the arid southwest. In
         Land and the pursuit of happiness, ed. E. Lenz and A. LeBel,
         pp. 11–20. Western Humanities Center, University of Cali-
         fornia, Los Angeles Extension.
Fontana, B.L., ed.
 1965   *An Englishman's Arizona: the ranching letters of Herbert R.
         Hislop, 1876–1878.* Overland Press, Tucson.
Ford Foundation
 1974   *A time to choose: America's energy future.* Energy Policy
         Project Series. Ballinger Publishing Co., Cambridge, Mass.
Frank, H.J.
 1975   Energy prospects and options for Arizona. *Arizona Review*
         24(2): 1–7.
Gavan, J.D., and J.A. Dixon
 1975   India: a perspective on the food situation. *Science* 188(4188):
         541–548.
Georgescu-Roegen, N.
 1971   *The entropy law and the economic process.* Harvard University
         Press, Cambridge.
 1975   Energy and economic myths. *Southern Economic Journal*
         41(3): 347–381.
Gerke, D., et al.
 1975   *Arizona State Water Plan, Phase I: inventory of resources and
         uses.* Arizona State Water Commission, Phoenix.
Gilliland, M.W.
 1975   Energy analysis and public policy. *Science* 189(4208): 1051–
         1056.

Gisser, M., and A. Mercado
  1972　Integration of the agricultural demand function for water and
        the hydrologic model of the Pecos Basin. *Water Resources
        Research* 8(6): 1373–1384.
Goodwin, G.
  1969　*The social organization of the western Apache.* University of
        Arizona Press, Tucson.
Graves, J., et al.
  1971　*The water hustlers.* Sierra Club, San Francisco.
Great Plains Committee
  1937　*The future of the Great Plains: message from the president of
        the United States transmitting the report of the Great Plains
        Committee under the title "The future of the Great Plains".*
        Committee on Agriculture, House of Representatives, 75th
        Congress, 1st Session, Document No. 144, Washington, D.C.
Green, D.E.
  1973　*Land of the underground rain: irrigation on the Texas High
        Plains, 1910–1970.* University of Texas Press, Austin.
Hackenberg, R.A.
  1955　*Economic and political change among the Gila River Pima
        Indians.* Report to the John Hay Whitney Foundation, March,
        1955. University of Arizona Bureau of Ethnic Research.
  1961a　*Aboriginal land use and occupancy of the Pima-Maricopa
        Indian Community.* Report to the U.S. Dept of Justice. Ari-
        zona State Museum, Tucson. [Xerox of typescript.]
  1961b　*A Pima bibliography.* Arizona State Museum, Tucson. [Xerox
        of typescript.]
  1964　*Aboriginal land use and occupany of the Papago Indians.*
        Arizona State Museum, Tucson. [Xerox of typescript.]
Hamilton, P.
  1884　*The resources of Arizona.* 3rd ed. A.L. Bancroft & Co., San
        Francisco.
Hansen, H.
  1969　Texas: a guide to the Lone Star State. Hastings House, New
        York.
Hardin, C.W.
  1952　The politics of agriculture: soil conservation and the struggle
        for power in rural America. Free Press, Glencoe, Ill.
Hardin, G.
  1975a　Gregg's law. *Bioscience* 25(7): 415.
  1975b　Life boat ethics: the case against helping the poor. *Congres-
        sional Digest* 54:205 plus.
Har-el, M.
  n.d.　*The ancient water supply system of Jerusalem.* Hebrew Uni-
        versity and Tel-Aviv University, Israel.

Harshbarger, J.W.
   1971   Groundwater resource evaluation and exploitation. Food and
           Agricultural Organization, United Nations, Seminario, Sobre
           Aguas Subterraneas, Granada, Spain, October 18–19, 1971.
Hartman, L.M., and D. Seastone
   1970   *Water transfers: economic efficiency and alternative institu-
           tions.* Published for Resources for the Future by Johns Hop-
           kins Press, Baltimore.
Hastings, J.R., and R.M. Turner
   1965   *The changing mile: an ecological study of vegetation change
           with time in the lower mile of an arid and semiarid region.*
           University of Arizona Press, Tucson.
Heady, E.O., H.C. Madsen, K.J. Nichol, and S.H. Hargrove
   1971   *Agricultural water demands.* National Water Commission
           Report NWC-F-72-031, November, 1971.
Hicks, J.D.
   1955   *The Populist revolt.* University of Minnesota Press, Minne-
           apolis.
High Plains Underground Water Conservation District, No. 1, Lubbock,
Texas.
   1970   Population to water abounds. *Cross Section* 16(8): 1–2.
Hirshleifer, J., J.C. De Haven, and J.W. Milliman
   1969   *Water supply: economics, technology and policy.* University of
           Chicago Press, Chicago.
Holcomb, W.H.
   1929   The development of the deep-well turbine pump. *Mechanical
           Engineering* 51(11): 833–836.
Holder, P.
   1970   *The hoe and the horse on the plains: a study of cultural devel-
           opment among North American Indians.* University of
           Nebraska Press, Lincoln.
Holdren, J.P.
   1973   Population and the American predicament: the case against
           complacency. *Daedalus* 102(4): 31–44.
Hood, O.P.
   1898   *Certain pumps and water lifts used in irrigation.* Water Sup-
           ply and Irrigation Papers of the U.S. Geological Survey, no.
           14, Washington, D.C.
Hornaday, W.T.
   1908   *Campfires on desert and lava.* Charles Scribner's Sons, New
           York.
Horowitz, M.M.
   1972   Ethnic boundary maintenance among pastoralists and farmers
           in the western Suden (Niger). *Journal of Asian and African
           Studies* (Leiden, Netherlands) 7(1–2): 105–114.

Howard, R.M.
  1959 Comments on the Indian's water supply at Gran Quivira National Monument. *El Palacio* 66:85–91.
Howe, C.W., et al.
  1971 *Future water demands: the impact of technological change, public policies, and changing market conditions on water use patterns of selected sectors of the United States economy, 1970–1990.* Resources for the Future, Inc., Washington, D.C.
Howe, J.W., and J.W. Sewell
  1975 What is morally right? *War on Hunger: A Report from the Agency for International Development* 9(6):1–5, 22–24.
Hubbert, M.K.
  1969 *Resources and man: a study and recommendations.* W.H. Freeman, San Francisco.
Hutson, W.P.
  1898 *Irrigation systems in Texas.* Water Supply and Irrigation Papers of the U.S. Geological Survey, no. 13, Washington, D.C.
Imperato, P.J.
  1972 Nomads of Niger. *Natural History* 81(10): 61–68, 78–79.
Irons, W.
  1972 Variations in economic organizations: a comparison of the pastoral Yomut and the Basseri. *Journal of Asian and African Studies* (Leiden, Netherlands) 7(1–2): 88–104.
Israelsen, O.W., and V.E. Hansen
  1962 *Irrigation principles and practices.* 3rd ed. John Wiley & Sons, New York.
Ives, R.L.
  1941 The origin of the Sonoita townsite, Sonora, Mexico. *American Antiquity* 7(1): 20–28.
  1950 The Sonoyta Oasis. *Journal of Geography* 49:1–12.
  1962 Kiss tanks. *Weather* 17:194–196.
  1964 *The Pinacate Region, Sonora, Mexico.* Occasional Papers of the California Academy of Sciences, no. 47. California Academy of Sciences, San Francisco.
James, G.W.
  1917a *Arizona the wonderland.* Page Co., Boston.
  1917b *Reclaiming the arid west.* Dodd, Mead & Co., New York.
James, L.D., ed.
  1974 *Man and water: the social sciences in management of water resources.* University Press of Kentucky, Lexington.
Jensen, C.W., and W.L. Track
  1973 *The Texas water plan and its institutional problems.* Technical Report no. 37, Texas Water Resources Institute, Texas A&M University.

Johnson, W.R.
   1973   Should the poor buy no growth? *Daedalus* 102(4): 165–190.
Jones, R.D.
   1969   An analysis of Papago communities, 1900–1920. Ph.D. dissertation, University of Arizona, Tucson.
Kazmann, R.G.
   1966   Safe yield in ground water development, reality or illusion? *Proceedings of American Society of Civil Engineers* 82(IR3): 1–12.
Kedar, Y.
   1957a   Ancient agriculture at Shivtah in the Negev. *Israel Exploration Journal* 7:178–179.
   1957b   Water and soil from the desert: some ancient cultural achievements in the central Negev. *Geographical Journal* 123:179–187.
   1959a   Ancient agricultural installations in the Negev mountains. *Studies in the Geography of Eretz-Israel* 1:124–125.
   1959b   The ancient agriculture in 'Avadat area. *Studies in the Geography of Eretz-Israel* 1:95–121.
   1962   Comparison of the ancient arroyo flood-irrigated agriculture in the southwestern United States with ancient agriculture in the Negev, Israel. *American Philosophical Society Yearbook*, pp. 567–568.
   1969a   Farming in arid zones: a study of ancient irrigated farming in the southwest of North America and in the Negev Mountains, Israel. Paper presented at the AAAS meeting, Colorado Springs, Colorado. Office of Arid Land Studies, University of Arizona, Tucson. [Xerox of typescript.]
   1969b   Limiting factors in land use in arid lands in ancient times. Presented to Arid Lands in a Changing World, an international conference sponsored by the AAAS Committee on Arid Lands, June 3–13, 1969, University of Arizona, Tucson.
Kelly, W.H.
   1963   *The Papago Indians of Arizona: a population and economic study.* Bureau of Ethnic Research, Department of Anthropology, University of Arizona, Tucson.
Kelso, M.M., W.E. Martin, and L.E. Mack
   1973   *Water supplies and economic growth in an arid environment: an Arizona case study.* University of Arizona Press, Tucson.
Kilcrease, A.T.
   1939   Ninety-five years of history of the Papago Indians. *Southwestern Monuments.* Supplement for April, pp. 297–310.
Kluckhohn, C., and D. Leighton
   1962   *The Navaho.* Doubleday & Co., New York.

Komie, E.E.
  1969 The changing role of the groundwater reservoir in arid lands.
        Presented to Arid Lands in a Changing World, an interna-
        tional conference sponsored by the AAAS Committee on Arid
        Lands, June 3–13, 1969, University of Arizona, Tucson.
Kunin, V.V.
  1957 Conditions of the formation of underground water in deserts.
        *International Association of Scientific Hydrology, General
        Assembly, Toronto, Publication 44* 2:502–516.
Lapp, R.
  1973 *The logarithmic century.* Prentice-Hall, Englewood Cliffs, N.J.
Larson, P.
  1970 *Deserts of North America.* Prentice-Hall, Englewood Cliffs,
        N.J.
McClain, C.
  1975 Arizona's water mismanagement threatens disaster: wild,
        wooly—and dry west? *Tucson Daily Citizen*, April 18, p. 27.
MacDougal, D.T.
  1908 Across Papaguería. *Plant World* 11(5): 93–131.
McGee, W.J.
  1895 The beginnings of agriculture. *American Anthropology*
        8:350–375.
  1898 Papaguería. *National Geographic Magazine* 9(8).
  1906 Desert thirst as a disease. *Interstate Medical Journal* 13:279–
        300.
  1971 *Seriland.* Rio Grande Press, Glorieta, N.M.
McGinnies, W.G., B.J. Goldman, and P. Paylore, eds.
  1968 *Deserts of the world: an appraisal of research into their physi-
        cal and biological environments.* University of Arizona Press,
        Tucson.
  1971 *Food, fiber, and the arid lands.* University of Arizona Press,
        Tucson.
McKean, R.N.
  1973 Growth vs. no growth: an evaluation. *Daedalus* 102(4): 207–
        229.
McMurtry, L.
  1968 *In a narrow grave: essays on Texas.* Encino Press, Austin.
MacPhail, D.D., ed.
  1972 The High Plains: problems of semiarid environments. Con-
        tribution No. 15 of the Committee on Desert and Arid Zones
        Research, Southwestern and Rocky Mountain Division, AAAS,
        Colorado State University, Fort Collins.
Mails, T.E.
  1972 *The mystic warriors of the plains.* Doubleday & Co., Garden
        City, N.Y.

Makhijani, A., and A. Poole
  1975   *Energy and agriculture in the third world*. Ballinger, Cambridge.
Malde, H.E.
  1964   Environment and man in arid America. *Science* 145(3628): 123–129.
Manje, J.M.
  1954   *Luz de tierra incognita: unknown Arizona and Sonora, 1693–1721*. Arizona Silhouettes, Tucson.
Mann, J.F.
  1961   Factors affecting safe yield of groundwater basins. *Journal of Irrigation and Drainage Division, American Society of Civil Engineers* 3(2948): 63–69.
Mark, A.K.
  1960   Descriptions of and variables relating to ecological change in the history of the Papago Indian population. M.A. thesis, University of Arizona, Tucson.
Martin, P.S.
  1963   *The last ten thousand years: a fossil pollen record of the American Southwest*. University of Arizona Press, Tucson.
Meadows, D.H., et al.
  1972   *The limits to growth: a report for the Club of Rome's project on the predicament of mankind*. Universe Books, New York.
Meinzer, O.E.
  1934   The history and development of groundwater hydrology. *Journal of Washington Academy of Science* 24:6–32.
Mishan, E.J.
  1973   Ills, bads and disamenities: the wages of growth. *Daedalus* 102(4): 633–688.
Moss, F.E.
  1967   *The water crisis*. Frederick A. Praeger, Publishers, New York.
Murphy, E.C.
  1897   *Windmills for irrigation*. Water Supply and Irrigation Papers of the U.S. Geological Survey, no. 8, Washington, D.C.
Murphy, L.
  1914   The farmer's wife. *Texas Department of Agriculture Bulletin No. 35, Proceedings of the Third Meeting, Texas State Farmers' Institute, 1913*, Austin, pp. 46–48.
National Academy of Sciences
  1974   *More water for arid lands: promising technologies and research opportunities*. National Academy of Sciences, Washington, D.C.
Newell, F.H.
  1902   *Irrigation in the United States*. Thomas Y. Crowell, New York.

Nir, D.
  1974 *The semi-arid world: man on the fringe of the desert.* Longman, New York.
Nulty, L.
  1972 *The green revolution in west Pakistan: implications of technological change.* Praeger Publishers, New York.
Odum, H.T.
  1970 Energy value of water resources. *Proceedings, Nineteenth Southern Water Resources and Pollution Control Conference,* April, pp. 56–64.
  1971 *Energy, the environment and society.* John Wiley & Sons, New York.
Olson, M.
  1973 Introduction. *Daedalus* 102(4): 1–14.
Olson, M., H.H. Landsberg, and J.L. Fisher
  1973 Epilogue. *Daedalus* 102(4): 229–241.
O'Neill, G.K.
  1974a The colonization of space. *Physics Today* 27(September): 32–40.
  1974b A Lagrangian community? *Nature* 250:636.
Osborn, J.E., and W.C. McCray
  1972 *The structure of the High Plains economy.* Department of Agricultural Economics, Texas Tech University, Lubbock.
Paddock, W.C.
  1970 How green is the green revolution? *Bioscience* 20(16): 897–902.
Padfield, H., and C.L. Smith
  1968 Water and culture: new decision rules for old institutions. *Rocky Mountain Social Science Journal* 5(2): 23–32.
Peckham, R.C., et al.
  1967 *Additional technical papers on selected aspects of the preliminary Texas water plan.* Texas Water Development Board, Report No. 38, Austin.
Peden, D.G., et al.
  1974 The trophic ecology of *Bison Bison L.* on shortgrass plains. *Journal of Applied Ecology* 11:489–498.
Petersen, W.
  1975 A demographer's view of prehistoric demography. *Current Anthropology* 16(2): 227–245.
Pfefferkorn, I.
  1949 *Sonora: a description of the province.* University of New Mexico Press, Albuquerque.
Pimental, D., et al.
  1973 Food production and the energy crisis. *Science* 182(4111): 443–449.

1975　Energy and land constraints in food production. *Science*
190(4216): 754–761.

Poleman, T.T.
1975　World food: a perspective. *Science* 188(4188): 510–518.

Potter, D.M.
1954　*People of plenty: economic abundance and the American
character*. University of Chicago Press, Chicago.

Powell, J.W.
1962　*Report on the lands of the arid region of the United States:
with a more detailed account of the lands of Utah*. Harvard
University Press, Cambridge.

Rappaport, R.A.
1967　*Pigs for the ancestors: ritual in the ecology of a New Guinea
people*. Yale University Press, New Haven.

Rayner, F.A.
1972　Groundwater management on the High Plains of Texas.
Groundwater 10(5): 12–17.

Ressler, J.Q.
1966　Spanish mission water systems, northwest frontier of New
Spain. M.A. Thesis, University of Arizona, Tucson.

Reynold, Jr., E.C.
1910　Irrigation pumping. *Electrical World* 55(17): 1064–1066.

Reynolds, J.
1970　*Windmills and watermills*. Praeger Publishers, New York.

Richards, C.E.
1964　*DineBitoh: Navajo water use*. Comparative Studies of Cultural Change, Department of Anthropology, Cornell University, Ithaca, N.Y.

Ridgeway, J.
1973　*The last play: the struggle to monopolize the world's energy
resources*. E.P. Dutton, New York.

Roberts, M.J.
1973　On reforming economic growth. *Daedalus* 102(4): 119–138.

Ross, C.P.
1923　*The lower Gila region, Arizona*. U.S. Geological Survey Water
Resources Paper No. 498, Washington, D.C.

Rund, N.H., H. Siegel, and E.G. Rumley
1968　*Demographic and socio-cultural characteristics: Papago Indian
reservations, Arizona*. Health Program Systems Center, Division of Indian Health Services, Public Health Service, Tucson.

Russell, F.
1908　The Pima Indians. In *Annual Report, Bureau of Ethnology*,
26:3–390, Washington, D.C.

Ruxton, G.F.A.
  1847   *Adventures in Mexico and the Rocky Mountains.* J. Murray, London.
Ryder, N.B.
  1973   Two cheers for ZPG. *Daedalus* 102(4): 45–62.
Safrany, D.R.
  1974   Nitrogen fixation. *Scientific American* 231:64–70.
Sageser, A.B.
  1967   Windmill and pump irrigation on the Great Plains, 1890–1910. *Nebraska History* 48:107–118.
Salzman, P.C.
  1972   Multi-resource nomadism in Iranian Baluchistan. *Journal of Asian and African Studies* (Leiden, Netherlands) 7(1–2): 60–68.
Sauer, C.O.
  1934   *The distribution of aboriginal tribes and languages in north-western Mexico.* Ibero-Americana: 5. University of California Press, Berkeley.
  1935   *The aboriginal population of northwestern Mexico.* Ibero-Americana: 10. University of California Press, Berkeley.
  1969   *Agricultural origins and dispersals: the domestication of animals and foodstuffs.* M.I.T. Press, Cambridge.
Saxton, D., and L. Saxton
  1973   *O'otham Hohoók Aágitha: legends and lore of the Papago and Pima Indians.* University of Arizona Press, Tucson.
Shaw, A.M.
  1968   *Pima Indian legends.* University of Arizona Press, Tucson.
  1974   *A Pima past.* University of Arizona Press, Tucson.
Sherbrooke, W., and P. Paylore
  1973   *World desertification: cause and effect.* Office of Arid Lands Studies, University of Arizona, Tucson.
Shreve, F.
  1951   *The vegetation of the Sonoran desert.* Publication 591, Carnegie Institution of Washington, Washington, D.C.
Simons, M.
  1967   *Deserts: the problem of water in arid lands.* Oxford University Press, London.
Simpson, E.S.
  1967   A general summary of the state of research on groundwater hydrology in desert environments. In *Deserts of the world: an appraisal of research into their physical and biological environments*, ed. B.J. Goldman and P. Paylore, pp. 727–746. University of Arizona Press, 1967.

Smith, C.L., and H.I. Padfield
   1969   Land, water and social institutions. In *Arid lands in perspec-
           tive*, ed. W.G. McGinnies, B.J. Goldman, and P. Paylore, pp.
           325–336. University of Arizona Press, Tucson.
Smith, H.N.
   1950   *Virgin land: the American West as symbol and myth*. Harvard
           University Press, Cambridge.
Smythe, W.E.
   1911   *Conquest of arid America*. Macmillan, New York.
Spicer, E.H.
   1962   *Cycles of conquest: the impact of Spain, Mexico, and the
           United States on the Indians of the southwest, 1533–1960*.
           University of Arizona Press, Tucson.
Spicer, E.H., ed.
   1952   *Human problems in technological change: a casebook*. Russell
           Sage Foundation, New York.
Spooner, B.
   1972   The status of nomadism as a cultural phenomenon in the
           Middle East. *Journal of Asian and African Studies* (Leiden,
           Netherlands) 7(1–2): 122–131.
Sprague, G.F.
   1975   Agriculture in China. *Science* 188(4188): 549–556.
Stakman, E.C., R. Bradfield, and P.C. Mangelsdorf
   1967   *Campaigns against hunger*. Harvard University Press, Cam-
           bridge.
Stamp, L.D.
   1961   *A history of land use in arid regions*. UNESCO, Nancy,
           France.
Steinhart, J.S., and C.E. Steinhart
   1974   Energy use in the U.S. food system. *Science* 184(4134): 307–
           316.
Steward, J.H., ed.
   1955   *Irrigation civilizations: a comparative study*. Pan American
           Union, Washington, D.C.
Stewart, G.R., and M. Donnelly
   1943   Soil and water economy in the pueblo southwest. *Scientific
           Monthly* 56:31–44, 134–144.
Swidler, W.W.
   1972   Some demographic factors regulating the formation of flocks
           and camps among the Brahui of Baluchistan. *Journal of Asian
           and African Studies* (Leiden, Netherlands) 7(1–2): 69–75.
Tadmor, N.H., L. Shanan, and M. Evenari
   1960   The ancient desert agriculture of the Negev: VI. the ratio of
           catchment to cultivated area. *Ktavim* 10 (3–4): 193–206.

Thomas, H.E.
  1962 *Water and the Southwest—What Is the Future?* U.S. Geological Survey Circular No. 469.
Thomas, R.K.
  1953 *Papago land use west of the Papago Indian Reservation south of the Gila and the problem of Sand Papago identity.* Arizona State Museum, Tucson. [Copy.]
Thompson, L.M.
  1975 Weather variability, climatic change, and grain production. *Science* 188(4188): 535–540.
Toulouse, Jr., J.H.
  1945 Early water systems at Gran Quivira National Monument. *American Antiquity* 10:363–372.
Turnage, W.V., and A.L. Hinckley
  1938 Freezing weather in relation to plant distribution in the Sonoran desert. *Ecological Monographs* 8:530–550.
Turner, R.M.
  1963 Growth in four species of Sonoran desert trees. *Ecology* 44(4): 760–765.
Ucko, P.J., and G.W. Dimbleby, eds.
  1969 *The domestication and exploitation of plants and animals.* Aldine Publishing Co., Chicago.
Underhill, R.M.
  n.d.   *Papago Indians: articles.* Arizona State Museum, Tucson. [Typescript.]
  1938 *Singing for power: the song magic of the Papago Indians of southern Arizona.* University of California Press, Berkeley.
  1939 *Social organization of the Papago Indians.* Columbia University Press, New York.
University of California Food Task Force
  1974 *A hungry world: the challenge to agriculture.* University of California Press, 1974.
U.S. Department of the Interior
  1974 Westwide study: critical water problems facing the eleven western states. Report and executive summary. 2 pts. Denver, Colorado. [Draft.]
Vandertulip, J.J., L.B. Seward, and J.T. Thornhill
  1966 *Technical papers on selected aspects of the preliminary Texas water plan.* Report No. 31, Texas Water Development Board, Austin.
Wade, N.
  1975 International agricultural research. *Science* 188(4188): 585–588.

Walker, G.
  1969   *Miracle in moccasins: my forty years as a missionary to the*
         Indians of the Southwest. Phoenician Books, Phoenix.
Wallace, E., and E.A. Hoebel
  1952   *The Comanches: lords of the South Plains.* University of
         Oklahoma Press, Norman.
Wallace, E.S.
  1955   *The great reconnaissance: soldiers, artists, and scientists on*
         the frontier, 1848–1861. Little, Brown & Co., Boston.
Walsh, J.
  1975   U.S. agribusiness and agricultural trends. *Science* 188(4188):
         531–534.
Walter, H., and E. Stadelman
  1974   A new approach to the water relations of desert plants. In
         Desert biology, vol. 2, ed. G.W. Brown, Jr., pp. 214–311.
         Academic Press, New York.
Walters, H.
  1975   Difficult issues underlying food problems. *Science* 188(4188):
         524–530.
Webb, G.
  1959   *A Pima remembers.* University of Arizona Press, Tucson.
Webb, W.P.
  1931   *The Great Plains.* Ginn & Co., Boston.
  1964   *The great frontier.* University of Texas Press, Austin.
White, Jr., L.
  1967   *Medieval technology and social change.* Oxford University
         Press, New York.
Wiener, A.
  1972   *The role of water in development.* McGraw-Hill, New York.
William, C.H.
  1911   Electric energy from coal for irrigation farming in Colorado.
         Electrical World 58(14): 805–811.
Wittwer, S.H.
  1975   Food production: technology and the resource base. *Science*
         188(4188): 579–584.
Wolff, A.R.
  1890   *The windmill as a prime mover.* John Wiley & Sons, New
         York.
Woodbury, R.B.
  1960   The Hohokam canals at Pueblo Grande, Arizona. *American*
         Antiquity 26(2): 267–270.
  1961a Climatic change and prehistoric agriculture in the southwestern
         United States. Paper presented at Conference on Solar Varia-
         tion, Climatic Change and Related Geophysical Problems,

New York Academy of Sciences and American Meteorological Society, January 24–28, New York.

1961b A reappraisal of Hohokam irrigation. *American Anthropologist* 63:550–560.

Yarham, E.R.

1958 Singing sands: a strange concert heard in the desert. *UNESCO Courier* 6:26–27.

Zeckhauser, R.

1973 The risks of growth. *Daedalus* 102(4): 103–118.

Zubrow, E.B.W.

1971 Carrying capacity and dynamic equilibrium in the prehistoric southwest. *American Antiquity* 36(2): 127–138.

1974 *Population, contact, and climate in the New Mexican pueblos.* Anthropological Papers of the University of Arizona, no. 24. University of Arizona Press, Tucson.

Index